School Culture Development in China - Perceptions of Teachers and Principals

RIVER PUBLISHERS SERIES IN COMMUNICATIONS

Volume 6

Series Editor

Prof. Dr. Xiangyun Du
Aalborg University, Denmark

Nowadays, educational institutions are being challenged when professional competences and expertise become progressively more complex. This is mainly because problems are more technology-bounded, unstable and ill-defined with the involvement of various integrated issues. To solve these problems, it requires interdisciplinary knowledge, collaboration skills, innovative thinking among other competences. In order to facilitate students with the competences expected in professions, educational institutions worldwide are implementing innovations and changes in many aspects.

This book series includes a list of research projects that document innovation and change in education. The topics range from organizational change, curriculum design and innovation, pedagogy development, to the role of teaching staff in the educational change process, and quality issues, among others. A cross-cultural perspective is studied in this book series that includes two layers. First, research contexts in these books include different countries with various educational traditions, systems and societal backgrounds. Second, the impact of professional and institutional cultures such as engineering, medicine and health, and teachers' education are also taken into consideration in these research projects.

For a list of other books in this series, visit www.riverpublishers.com

School Culture Development in China - Perceptions of Teachers and Principals

Dr. Kai Yu

Beijing Normal University
China

Dr. Xiangyun Du

Aalborg University
Denmark

Dr. Xiaoju Duan

Aalborg University
Denmark

River Publishers
Aalborg

Published, sold and distributed by:
River Publishers
Niels Jernes Vej 10
9220 Aalborg Ø
Denmark

ISBN: 978-87-93102-66-8
978-87-93102-72-9 (Ebook)

Contents

1

School Culture Development in China

1.1 Educational Development in China

Education thrives in China because it is a top priority for the nation. Guan Zhong, an enlightened prime minister of the Spring and Autumn Period over 2,500 years ago, said, "For one year, nothing is more important than nurturing grain; for ten years, nothing is more important than nurturing trees; for a lifetime, nothing is more important than nurturing men." Today, Chinese people recognize the importance of an educated work force for economic growth, and they understand that investing in the education system makes their economy globally competitive. "At the end of 2007, the population of students in primary and middle schools reached over 200 million, and the number of teachers exceeded 9 million. There are more than 320,000 primary schools, 5900 secondary schools and 1600 general high schools" (Zhang et al., 2010). For a figure illustrating the Chinese education system, see Appendix 1.

Total national funding for education has on average increased 20 percent annually since 1993. Education funding reached 548 billion yuan in 2002, over five times as much as in 1993 (Tsang, 1996), and 2.2 trillionyuan in 2013 (Business Daily Update), meeting the goal set 20 years earlier. With this heavy investment in education, China sought to provide its population with greater access to elementary and secondary education. Two decades of reform and development have created an entirely new educational landscape in China. Over 98 percent of populated areas have realized nine-year compulsory education. All students in rural areas enjoy free compulsory education - free tuition, free textbooks, free of miscellaneous fees. The enrollment rate for high school reached 59.2 percent in 2006, on track to reach 90 percent by 2020, according to the *National Outline for Medium- and Long-Term Educational Reform and Development (2010–2020)*. Central and local governments are responsible, respectively, for guaranteeing and promoting the development of K-12 education. Private education enrolls about 5 percent of total primary, middle, and high school students. Most of the K-12 teachers are highly motivated. School building construction and infrastructure have improved a great deal. Enrollment in higher education has also increased dramatically. "Since 1999, institutions of higher learning have enrolled more and more students every year, so that by 2002, there were 16 million students in different types of higher education institutions. Of these, some 9.03 million were attending regular colleges and universities" (Li, 2004). The number of new university enrollments each year has quintupled, rising from 1 million students in 1997 to more than 6.85 million in 2013. This expansion is unprecedented worldwide, and university enrollment in China is now the largest in the world.

Some pressing challenges still remain, however. Although K-12 education in China has made significant advances, its overall development has been quite uneven and imbalanced. Obvious gaps exist among different geographic areas (Wang and Zhao, 2011), schools (Wang, 2001), and student groups (Zhao, 2009). The education system can neither respond effectively to the needs of rapid socio-economic development nor satisfy the aspirations of the general population for balanced basic education. The quality of K-12 education cannot meet the increasing demand for high-quality education from parents, especially in the era of the one-child policy (Tan, 2012).

In contrast to its K-12 education, Chinese higher education has faced harsh criticism. Many Chinese scathingly complain that their higher education system kills independent thought and creativity. Students have voted with their feet. In 2009, 10 percent of the more than 10 million potential college student candidates elected not to take the National College Entrance Examination (NCEE) (Carducci, 2012). The number of students taking the NCEE is expected to continue to drop in the coming years. According to *China Daily*, in 2007, 144,000 Chinese had studied abroad, while 44,000 among them (29.88 percent), had come back.

Since 1996 curriculum reform has been an important element of China's basic education reform effort. The Ministry of Education (MOE) has developed national curriculum standards for all core curricular areas. National curricula constitute 80 percent of curriculum workload, provincial curricula 15 percent, and school curricula 5 percent. The MOE sets guidelines for curriculum design done at the provincial level, yet does not interfere with its autonomy. This also makes it possible for each school may set its own school-based curriculum.

In 2001, over 550 counties piloted the new curricula in their compulsory grades, usually grades one through nine. By 2005, all primary and middle schools in all counties throughout the country's 31 provinces, municipalities, and autonomous regions had implemented the new curriculum. In 2004, four provinces piloted the curricula at their high schools. In 2007, 16 provinces total followed the new high school curriculum (Ryan, 2011). Most provinces were advised to implement the new high school curriculum by 2008. "The underlying concept of the current round of curriculum reform is to enhance the students' initiative to learn in the course of school education such that they do not just learn the 'knowledge', but also 'learn how to learn', acquire creativity and abilities for lifelong learning, and have the ability to apply knowledge learned" (Luo, 2011). Because of its importance and its close connection with the National College Entrance Examination, high school education has become the focus of the current round of reforms of China's basic education.

One requirement of new curriculum reforms is an increase in the professional standards for China's teachers. According to the Ministry of Education of China October, 2007, 240,000 teachers still did not meet the minimum requirement for academic degrees, and 450,000 teachers were not licensed. However, China has decided to raise academic standards for teachers even further: from a normal school degree to a three-year college degree for primary school teachers, and from a three-year college degree to an undergraduate degree for middle school teachers. Central and provincial governments are working together to encourage teachers to rotate among schools in order to achieve educational equality in many aspects.

Classroom learning and class life are the core elements of students' school life. The new generation of students is empowered to question, to challenge, to think for itself, and to participate equally in classroom life (Yang, 2012). The Ministry of Education has campaigned heavily for teaching and learning reform, including effective usage of interactive learning, promoting research-oriented study and teamwork, focusing on how to learn rather than what to learn, etc. To remove obstacles to teaching and learning reform, the central government also plans to reform the existing evaluation system in K-12 schools. For instance, it has replaced the hundred-mark system with a five-point grade and requires more encouragement to be given to students; it has also sought continuous innovation in the National College Entrance Examination system. For example, a list of elite universities were permitted to designate their own entrance examination and admission standards since 2003, and in a pilot program starting from 2010 the Ministry of Education allowed about 100 prestigious high schools to recommend a certain number of students to Peking and Tsinghua Universities since 2011 (Bu & Li, 2013).

1.2 Basic Education in Beijing

Beijing has a wealth of human resources like no other city in China. Owing to the continuous efforts of the local government, all children in Beijing are entitled access to basic education, and the quality of the education in terms of educational level and competences gained has been significantly improved. Appendix 2 shows the indexes that reflect the universalized qualities.

There has been a significant increase in the number of people with access to high-level education. The extent of basic education universalization can be charted by the net enrollment rate of primary school age children, the advancement rate of primary school graduates, the gross middle school enrollment rate for the age group, and the advancement rate of middle school graduates (Xin & Kang, 2012). Appendix 3 shows changes and trends in the universalization of Beijing's basic education in recent years.

According to Donald J. Treiman of the University of California, Los Angeles, "the upward trend [in educational and human capital attainment] was mainly fueled by the expansion of primary and lower middle education, resulting in a nearly complete gender convergence, as women went from virtually no schooling at the beginning of the [twentieth] century to nearly universal primary matriculation by the end of the century" (Treiman, 2013).

The two decades of K-12 education expansion in Beijing have seen continuously striving toward universalizing basic education and bringing about educational equality among different areas and population groups. Appendix 4 illustrates the number of primary students and the schools and classes by district and grade in Beijing.

Hu et al. (2009) noted that educational expenditure in Beijing is far above the national average: "national educational expenditure per capita is 931.54 RMB in 2003 as compared with 491.48 RMB in 2000, with an increase of 190 percent. However, the figure in Beijing is 1915.41 RMB (2000) and 3348.24 RMB (2003), increasing by 175 percent". The authors found that schools in inner city districts had lower educational efficiency (the ratio of educational inputs such as funding to outputs such as the number of graduates and their academic achievements) despite having better education inputs. According to their study, there are four conceivable factors that can be related to this issue: (1) The schools in the main city zone are more complex, and the gap in educational quality between schools can be very large, even within a single district; (2) teachers are less satisfied with their salaries; (3) large class size hinders improvements in efficiency; and (4) more schooling hours in the main city zone compared with the suburbs results in more student dissatisfaction.

The shift away from the traditional orientation of basic education to an emphasis on "learning how to learn" in Beijing requires teachers to be active participants in order to be effective. Appendix 5 shows the number of primary teachers and staff members by district in Beijing.

Beijing's current push for progress in its K-12 schools reflects widespread concern about its ability to rise to the level of national and global competition. Despite scattered inspiring educational reforms at some schools, the essential nature of schools in Beijing remains unchanged, a situation that has long plagued Beijing's education reformers. Therefore, a new path toward change needs to be found.

1.3 The School Culture Movement in Beijing

School culture reflects the norms; goals; values; interpersonal relationships; teaching, learning, and leadership practices; and organizational structures that comprise school life. It involves people's assumptions about and perceptions of school life, the social, emotional, ethical, and civic as well as intellectual aspects of learning and school improvement. Beijing's initiatives related to school culture were situated within the national campaign for the connotative and sustainable development of education, which was influenced by a series of important educational events.

On December 7, 2007, Beijing's municipal government announced the launchof the five-year Primary School Standardization Construction Program. The eighth article of the official founding document promoted connotative development and quality education in the city's K-12 schools, and the fourteenth article proposed a resource reassignment system, promising to advance the characteristic and unique development of the schools in Beijing's jurisdiction. One year later, in December 2008, with the help of a group of scholars from Beijing Normal University, the local educational authorities of Haidian District initiated its School Culture Program, starting with its eight primary schools.

Haidian is an important district in Beijing in terms of education and research. It has a history of successfully implementing academic innovations and of high academic achievements. As a famous saying goes, "Go to Beijing if you want the finest element of Chinese education, and go to Haidian if you want the best of the best." In 2011, the local educational authorities of Haidian District decided to focus on school culture in all primary schools within its jurisdiction in order to finish the Primary School Standardization Construction Program with a satisfactory completeness.

Other large districts jumped into the school culture campaign. In March 2010, the local educational authorities of Shijingshan District worked together with Beijing Normal University to turn this traditionally heavy industrial area into a brand-new educational experimentation area. This project was ambitious with the support of a large account of funding from the local government. The chief of the Shijingshan Commission of Education said in her address, "We used to feel proud of producing high-quality steel and came to realize that we have to bring up a high-quality work force, to rise to the competition we are faced with. We want to be a cultural powerhouse, rather than an industrial base, for this city. Let's start with our education." Following Haidian, Shijingshan chose six schools to pilot the school culture project. In 2012, another big education district, Chaoyang, decided to catch up with its peers in the school culture race. Chaoyang focused its school culture project in three areas: (1) institutional culture, i.e., democratic management and decision-making process, etc.; (2) spiritual culture, i.e., school motto, banner, anthem, etc.; (3) student behavior culture, i.e., student clubs, after-school programs, arts, music, dancing, sports, etc.

At the municipal level, the Beijing educational authorities went further. They started an instructional leadership culture program in January 2012 in order to get principals involved in the school culture movement. Nine schools were chosen to pilot the experiment, including one school from Fengtai, four from Haidian, two from Chaoyang, one from Xicheng, and one from Dongcheng.

Then on July 11, 2012, Beijing's municipal government issued the *Outline for Moral Education in Beijing Elementary and Secondary Schools (2012–2015)*, which signaled an official citywide endorsement of the school culture movement in its K-12 education. This was the tipping point for the school culture movement as Beijing planned to set up 500 exemplary schools that would be pioneers in the field of school culture by the end of 2015.

The school culture movement in Beijing was interwoven with the city's desire to set up schools that carried with them a brand name like recognition throughout China. Beijing is the most significant historic city in China, so the local educational authorities decided to start with the oldest schools with the hope of setting an example for other schools in Beijing. On August 21, the Beijing Commission of Education selected 33 century-old schools, including 11 schools from Dongcheng District, 13 from Xicheng, 2 from Haidian, and the other 7 schools from Tongzhou, Shunyi, Changping, Daxing, Huairou, Miyun, and Yanqing. The chosen schools met all of the following four standards: (1) they had over 100 years of uninterrupted schooling; (2) they had a good compilation of school history, including a school journal, yearbook, and chronicle; (3) they had distinct school traditions and characteristics; and (4) they had high academic reputations.

On October 25, 2012, the Beijing Commission of Education and the Primary School Education Association of the Beijing Society of Education launched the Forum on Cultural Inheritance and Innovation of Century-old Primary Schools in Beijing, No. 1 Experimental Primary School. More than 300 primary school principals attended the forum. Luo Jie, Deputy Director of the Beijing Commission of Education, stressed in his concluding address that the century-old primary schools should exploit their cultural traditions and define their school culture by a group of core values. Luo also urged that all Beijing primary principals improve their schools by accurately assessing their schools' educational position, i.e., the performance level of their school, the direction their school is going in, and their student population in order to better serve their target school populations. On January 5, 2013, the Primary School Education Association of the Beijing Society of Education held its third board meeting. The meeting summarized the lessons of the century-old primary school campaign, and board members decided to focus the 2013 theme of the association on school culture and school improvement.

High schools, traditionally overwhelmed by the NCEE (Gaokao), did not want to lag behind this time. On December 21, 2012, the High School Education Association of the Beijing Society of Education and all elite high schools of Beijing, including Beijing No. 4 High School, Jingshan School, Beijing No. 80 High School, Tsinghua University High School, and Beijing Normal University High School, held the first Xiangshan Mountain Forum. The theme of the forum was school culture and school spirit, which echoed the appeal of the Beijing municipal government for the school culture movement.

The Beijing Commission of Education decided to fuel the schools' fervor for school culture and sought broader and faster progress. It issued an official notification on the selection of exemplary schools in school culture on October 10, 2013. Beijing announced that it would choose 100 exemplary schools in 2013 and then another 150 in 2014. Finally, it will seek to set up 500 exemplary schools before the end of 2015. Appendix 6 shows the 2013 quota of exemplary schools by district.

The selection of exemplary schools was not the final goal of the Beijing Commission of Education. It had a more ambitious plan to push forward the school culture movement in its K-12 education system and did so by establishing an evaluating indicator system for school culture, which also serves as a guideline for schools to follow if they wanted to make their mark this field. The evaluating system is consisted of a list of indicators including Understanding and Ideas of School Culture (with five sub-indicators: accuracy, integrity, integration, feasibility, and consensus), Spiritual Culture (with four sub-indicators: core values, goals of school development, goals of student development, and school motto), Institutional Culture (with four sub-indicators: legal awareness, principal accountability system, check and balance system, parental involvement, and community cooperation), Operating Culture (with four sub-indicators: management, curriculum and instruction, faculty development, student development, and teacher-student interaction), Environment Culture (with three sub-indicators: natural environment, school climate, and usage of space), Historic Culture, and Cultural Characteristics. Details of the evaluating indicator system for school culture are shown in Appendix 7.

On October 16, 2013, one week after the announcement of the selection of exemplary schools, the Beijing Commission of Education published *Suggestions on Promotion of Primary Education Quality*. The *suggestions* placed heavy emphasis on school culture. The recommendations included encouraging formulation of school by laws, improving institutional culture by way of a system of checks and balances, boosting school climate, setting up exemplary schools and child-friendly schools, etc.

In summary, it is no accident that different levels of local government in Beijing have chosen school culture as a path to improve their city's K-12 education. With the advancement of Primary School Standardization Construction Program and the Exemplary Middle School Program, K-12 schools in Beijing have experienced significant updates in terms of school buildings, information technology, teachers' professional development, etc. However, the rising middle class in the Beijing metropolitan area held higher expectations for their children's education. School choice became a serious social problem in Beijing. To bridge the gap between high expectations and scarcity of elite schools, Beijing's local educational authorities had no other choice than focusing on connotative development of the schools in its jurisdiction. It was estimated that beginning in 2010 Beijing would see a new wave of baby boomers reaching school age. School culture provides an effective mechanism for educational leaders to do more with less and to boost general educational quality within a short period of time. That is to say, with a focus on school culture, every school in Beijing can shine in a different way and meet the education needs of its target population.

1.4 Defining this Book

Within the national campaign for the innovative and sustainable development of education, local governments in Beijing Municipality have encouraged and facilitated educational innovation towards a better school culture. In the past three years, various initiatives have been undertaken by a select group of schools in order to develop strategies and enhance action towards change. The first phase of the campaign turned out to be a trial for exploring methods and gaining experiences. In 2013, a longitudinal research project was initiated, which has the support and collaboration of the Beijing Educational Commission and researchers and collaborators from Beijing Normal University. The overall aims of the research project are to:

- Gain scientific knowledge about the experiences and outcomes of the first phase of the school culture development campaign within Beijing Municipality.
- Identify challenges and possibilities of further developing school culture in Beijing Municipality and a broader Chinese context.
- Pinpoint needs and propose recommendations for further educational development towards better school culture.
- Contribute to school culture development practices and research internationally.

With these overall aims, a pilot study was initiated in 2013 to gain an overview of the current degree of school culture development by investigating the perceptions of actors of change. Although the Beijing Commission set up an evaluation system to help them choose and identify the exemplary schools in the field of school culture and used visible results (environment, school brochure, pamphlet, coursework, etc.) to evaluate school culture, there was a general lack of knowledge of insiders' views, i.e., what the actors thought of the school culture they work in. Therefore, teachers and principals have been chosen as the target group for this pilot study because they are regarded as the key actors in leading and implementing educational innovation and making cultural changes in schools (Deal and Peterson, 1999, 2009; Fullan 2001, 2007; Preble and Gordon, 2011). Our review of the literature on school culture in an international context identified an agreement on several core issues, namely the involvement of staff in the change process, consistency of change, adaptability of the organization to sustain change, and the mission of the school (for details see Chapter 2). Therefore, the pilot study was set up to address the following research questions: What are the perceptions of teachers and principals on school culture development in terms of their involvement, consistency of the cultural change, adaptability of the school, and mission of the school in the educational change process? Do teachers and principals have common perceptions of the school culture at their schools, and if so, in which ways?

A survey-based investigation was begun as a pilot study in 2013. The Denison Organizational Culture Survey (with the permission from Denison Consulting) was employed as the major data generation instrument. The survey was conducted in Chinese. Minor revisions were made to the original survey in order to make the translation more closely fit the Chinese-language context and to better suit the school context in China (for the English version of the survey, see Appendix 8, 9, and 10, and for more information about the survey, see Chapter 3). The investigation was conducted at 37 schools in four different districts in Beijing Municipality. 2066 teachers and principals participated in the investigation, and 1992 valid responses were received. Results of the study are reported and discussed in the remaining chapters of the book.

1.5 Organization of the Book

Chapter 1 maps out the landscape of Chinese education: traditions, funding, education system, curriculum reform, grading and assessment system for academic achievement etc. It also uses tables to describe various aspects of basic education in Beijing: its education and human resources indicators, its education scale and schooling population disaggregated by districts and grades, etc. The final section of the chapter explains why and how different levels of local educational authorities in Beijing have focused on school culture as a means to respond to the increasing demand from the middle class for high-quality K-12 education.

Chapter 2 presents and discusses the basic concepts of culture and school culture. A literature review on school culture studies gives an overview of the characteristics of a good school culture as well as the practices that are needed to construct a great school culture. An organizational perspective is taken in this book in order to study school culture in China from the perspectives of principals and teachers.

Chapter 3 presents the methodology of this study, including an introduction to the surveys as well as explanation of data generation and analysis. Also reported in this chapter is respondents' descriptive data.

Chapters 4–7 report the results of this study, structured according to the four themes of teachers' and principals' perceptions of the school culture traits: involvement (in chapter 4), consistency (in chapter 5), adaptability (in chapter 6), and mission (in chapter 7).

Chapter 8 begins with a summary of the results of the study and discusses these results in relation to the overall research questions and relevant literature as well as contexts introduced in chapters 1 and 2. This is followed by a reflection upon the pilot study project as well as recommendations for further steps toward school culture development in China.

2

Conceptual Understanding of School Culture

In this chapter we present and discuss the basic concepts of culture and in particular, school culture. After reviewing diverse perspectives on the conceptual understanding of culture, this study takes its standing point from a complex perspective that allows us to see culture in general, and more specifically school culture, as dynamic in nature and as a perpetual process of change. The literature review of existing studies on school culture also gives an overview of the characteristics of a good school culture as well as the practices involved in constructing a great school culture. After this overview, research on school culture in China is presented. This chapter ends with an introduction to the research choices that were made in this study such as why an organizational perspective was used, how the approach to studying school culture in China was employed, and why principals and teachers were the primary focus in this study.

2.1 Understanding Culture

The notion of culture is a term often used and discussed both in scholarly works and in the everyday lives of people as a way to explain and to interpret the social reality. Culture among many other areas, is an important notion in the analysis of any social science or humanities (Schoen and Teddlie, 2008). A cataract of literature has been published on culture from diverse perspectives arranging from education, linguistics, language studies, communication, anthropology, psychology, organization and management studies, among others. Accordingly, definitions of culture vary depending on where it is studied and the angles and perspectives used by the research to study it. The following sample definitions capture the diverse ways culture can be understood.

- The *American heritage dictionary of the English language* (1992: 454) defines culture as "1. a. The totality of socially transmitted behavior patterns, arts, beliefs, institutions, and all other products of human work and thought. b. These patterns, traits, and products considered as the expression of a particular period, class, community, or population: Edwardian culture, Japanese culture, the culture of poverty. c. These patterns traits and products considered with respect to a particular category, such as a field, subject or mode of expression: religious culture in the Middle Ages; musical culture: oral culture. 2. Intellectual and artistic activity, and the works produced by it. 3. a. Development of the intellect through training or education. b. Enlightenment resulting from such training or education. 4. A high degree of taste and refinement formed by aesthetic and intellectual training."
- "Culture consists of patterns, explicit and implicit, of and for behavior acquired and transmitted by symbols, constituting the distinctive achievements of human groups, including their embodiments in artifacts; the essential core of culture consists of traditional (i.e. historically derived and selected) ideas and especially their attached values; culture systems may, on the one hand, be considered as products of action, in the other as conditioning elements of further action" (Kroeber and Kluckhohn, 1952: 181).
- "Culture, conceived as a system of competence shared in its broad design and deeper principles, and varying between individuals in its specificities, is then not all of what an individual knows and thinks and feels about his world. It is his theory of what his fellows know, believe and mean, his theory of code being followed, the game being played, in the society into which he was born" (Keesing, 1981: 58).
- Culture refers to "the innumerable aspects of life. To most anthropologists, culture encompasses the behaviors, beliefs, and attitudes that are characteristic of a particular society or population" (Ember and Ember, 1981: 25).
- Culture "consists of the habits and tendencies to act in certain ways, but not the actions themselves. It is the language patterns, values, attitudes, beliefs, customs, and thought patterns... not the things or behavior, but forms of things that people have in mind, their models for perceiving, relating and otherwise interpreting them" (Barnett, 1988: 102).
- "Culture consists of the stable, underlying social meanings that shape beliefs and behavior over time" (Deal and Peterson, 1990: 7).

A rich body of literature links the concept of culture to its organizational implementation. Martin (2002) provides a thorough literature review of the field of organizational culture and distinguishes between what organizational culture is and is not. She outlines three traditions of defining culture from an organizational perspective: functionalism, critical theory, and postmodernism. Martin also lists 12 definitions of culture, varying in terms of narrowness and depth of focus and criticizes how organizational researchers use and operationalize cultures in their studies in many different and often self-contradictory ways. She proposes using the definition of culture given by Smircich (1983) as it includes many manifestations of culture in a broad way. Smircich argues that a functionalist position assumes that culture is something an organization has, whereas an interpretive position views culture as a root metaphor (e.g., something that an organization is), while postmodernist position focuses on culture as a process largely defined by the communication of organizational life (1983: 13–14). Given these three approaches to defining culture, Smircich proposes culture be understood as "In a particular situation the set of meanings that evolves gives a group its own ethos, or distinctive character, which is expressed in patterns of belief (ideology), activity (norms and rituals), language and other symbolic forms through which organization members both create and sustain their view of the world and image of themselves in the world" (1983: 56).

Culture can also be defined according to its specific context. In everyday life, one uses the phrases "youth culture", "popular culture", "organizational culture," or refers to the culture of a particular group. Culture is often defined by distinguishing it from one culture or another. For example, culture can be categorized into big "C" and small "c", "high culture" and "low culture," "common culture" and "distinct culture" etc. (Steele and Suozzo, 1994). Culture can be a national culture referring to "something that some in the society share but not all" (Jensen, 2007), for example, Chinese culture, Danish culture, French culture, etc. With this complex conception of culture, a national culture can be seen as "something that some in the society share but not all," and new cultures develop "when individuals from diverse cultures interact in the society" (p.88).

Jensen (2007) summarizes two ways of defining and understanding culture: the descriptive concepts of culture and a complex concept of culture. From the descriptive concepts perspective, culture can be artifacts, beliefs and customs, or everything that is non-biological in a society, including behavior and concepts. It can also be about group identity, which expresses value and establishes stereotyping (Hecht et al., 2005). This approach tends to emphasize that everyone in a group shares the same culture and focus on collective identification in a group. However, it leaves a fixed image of the group and ignores the nature that the culture itself continues to shift and change. Thus, it risks obscuring the dynamic nature of culture and the diversity within the group (Hecht et al, 2005).

In contrast, from a complex concept perspective, theorists focus on processes and see culture as an active process of meaning-making. They focus on change, development, practice and procedures of culture (Hecht et al., 2005). Culture is seen as something that is temporal, emergent, unpredictable, and constantly changing and is seen as consisting of shared knowledge and meanings and values (Jensen, 2007). This view of culture also recognizes that not everyone in a nation may share these same elements of culture (Kahn, 1989; Jensen, 2007). This conception encourages a complex view of culture that considers events longitudinally instead of statically and captures the dynamic nature of culture. However, it may "miss the very elements that create the process (the structures) and neglect the purposive nature (function) of an activity" (Hecht et al, 2005, p. 57).

Agreement on the definition and understanding of culture has never been reached. Over time the definition of culture has shifted in meaning and has been contested in sharp debate (Rosaldo, 2006). In this book, we take our theoretical departure of understanding culture from a process perspective, focusing on the complex nature of culture because it permits us to emphasize the changes, developments, practices, and procedures of culture. It allows us not only to see what culture is, but also to better understand how culture operates in relation to educational practice.

2.2 School Culture

In the past decades, the concept of culture has been highly reflected and closely related to education and the teaching-learning process. Schools have distinct cultures, as Waller wrote in 1932 (cited from Deal and Peterson, 1999: 2–3): "Schools have a culture that is definitely their own. There are, in the school, complex rituals of personal relationships, as set of folkways, mores, and irrational sanctions, a morse code based upon them. There are traditions, and traditionalists waging their world-old battle against innovators."

In literature on school studies, two terms are often used without clear distinction – "school culture" and "school climate." Many authors tend to use them interchangeably (Hoy et al., 1991, Maslowski 2001) although some highlight their differences (Schoen and Teddlie 2008). In this study we use the term school culture in a broad sense which allows it to refer to school climate without paying special attention to the potential differences.

The notion of culture is considered an important factor in relation to education. Definitions of school culture are diverse, and as stated by Deal and Peterson (1999), there is no single universally agreed upon best definition of school culture. A review of the literature in the past two decades identifies several commonly addressed definitions:

- School culture is "a pattern of basic assumptions–invented, discovered, or developed by a given group as it learns to cope with problems... that has worked well enough to be considered valid and, therefore, to be taught to new members as the correct way to perceive, think, and feel in relation to those problems" (Schein, 1985: 9).
- School culture consists of "unwritten rules and traditions, norms and expectations that permeate everything: the way people act, how they dress, what they talk about, whether they seek out colleagues for help or don't, and how teachers feel about their work and their students" (Deal and Peterson, 1999: 2–3).
- School culture is "the way we do things around here" and "the basic assumptions, norms, and values and cultural artifacts of a school that are shared by school members, which influence their functioning at the school." It is therefore regarded as a holistic entity that pervades and influences people within a school (Maslowski, 2001: 8–9).

Many authors agree on the importance of school culture and its strong impact on students, teachers, administration, and principals. As Barth notes, "a school's culture has far more influence on life and learning in the school house than the state department of education, the superintendent, the school board, or even the principal can ever have" (2002: 7). Based on his empirical studies, Fullan (2005) suggests that successful schools often have a more demanding culture which encourages improving, promotes, and holds more hope for children than less successful schools. School culture development is also believed to be positively linked with school effectiveness (Van Houtte, 2011).

Deal and Peterson (2009: 12–13) summarize the functions and impacts of the culture has on a school: 1) culture fosters school effectiveness and productivity, 2) culture improves collegiality, collaboration, communication, and problem-solving practices, 3) culture promotes innovation and school improvement, 4) culture builds commitment and kindles motivation, 5) culture amplifies the energy and vitality of school staff, students and community, 6) culture focuses attention on what is important and valued.

Garcia and Dominguez (1997, from Garcia and Guerra, 2006: 105–106) summarize the following characteristics that the conceptualization of culture in an educational context reflects:

1. Culture provides the lens through which we view the world; it includes shared values, beliefs, perceptions, ideals and assumptions about life that guild specific behavior.
2. A distinguishing characteristic of cultural values is that they are shared by members of the group, rather than reflecting individual beliefs. While not all members of a culture ascribe to these values, these beliefs represent group tendencies.
3. Cultural values will persist, even though people who adhere to them may not express them consistently across time and place.
4. Culture is a dynamic process, likely to change over time and across generations. Group patterns of thought and behavior that are most likely to be transmitted intergenerationally include adaptive processes that promote group survival.
5. Cultural values guide people's behavior in unfamiliar settings: they provide the script that influences selection of behaviors and responses perceived to be appropriate for these settings.
6. Culture provides the basis for childhood experiences through which children are socialized (enculturated) to the norms, values, and traditions of their cultural group.
7. In educational contexts (formal and informal), culture influences each group's shared beliefs about and expectations for what children should be taught, how it should be taught, and by whom such instruction should be provided.

A rich body of literature on school improvement and change suggests how the cultures in the schools studied plays an important role in improving curriculum, teaching and learning methods, student achievement and professional development (Deal and Peterson, 1999, 2009; Fullan, 1998, 2001, 2007, Beaudoin and Taylor, 2004; Kruse and Louise, 2009; Muhammad, 2009; Carter, 2011; Preble and Gordon, 2011). Empirical evidence from these studies suggest that schools where improvement in norms, values and beliefs, can be observed, a sense of community, social trust among staff, and a shared commitment to school improvement can also be found. In contrast, when school cultures did not support or encourage reform, changes did not take places. Culture was a key factor in enhancing more effective practices (Deal and Peterson, 2009: 9).

What makes a "great school culture" has been studied. As summarized by Carter, "Great school cultures are explicit about what is valued, about what is truly good and about what they aim for. Through intentional practices and purposeful activity, they help the entire community strive in this direction" (2011: 18). Many authors have also made proposals for constructing a good school culture based on research-based practice.

Elbot and Fulton (2008: 13) propose employing Plato's three ways of knowing as a philosophy for building up school cultures. The first way of knowing is "the true", referring to academic contents and skills. It is also important for schools to go beyond the true by connecting knowledge to everyday life. The second way of knowing is "the good," which is about being virtuous: identifying with not only "me" but also with "we." It is about empathy, service to others, good citizenship, and character-building education. The third way of knowing is "the beautiful," which is about spirit, presence, poetry, and artistic sensibility. This way, learning goes beyond information and practical application to exploring the possibilities of being a human being. Based on the notions of the three ways of knowing, the authors suggest a framework consisting of eight essential aspects to be considered in the process of building up school cultures. The framework has named eight gateways for intentional school culture including: 1. Teaching, learning, and assessment 2. Relationships 3. Problem solving 4. Expectations, trust, and accountability 5. Voice 6. Physical environment 7. Markers, rituals, and transitions and 8. Leadership (chapter 4: 73–116).

Core norms and values are central to initiating, planning and implementing improvements. As Deal and Peterson (2009: 11) note, 1) a school with a strong, shared sense of mission is more likely to initiate improvement efforts, 2) norms of collegiality are related to collaborative planning and effective decision making, 3) cultures with a strong dedication to improvement are more likely to implement complex new instructional strategies, 4) schools improve best when small successes are recognized and celebrated through shared ceremonies commemorating both individual and group contributions.

Deal and Peterson (2009: 12) show how professional learning communications play an important role in reinforcing cultural elements in school success which include the following aspects: A shared sense of purpose, teacher involvement in decision making, collaborative work around instruction, norms of improvement, professional learning by staff, a sense of joint responsibility for student learning.

Carter (2011) proposes four stages necessary in creating a great school culture: 1) school culture begins to form in response to a need 2) school culture commits to certain founding principles 3) school culture begins to shape outcomes and 4) customs and habits are further embedded and improve outcomes.

Kruse and Louise (2009) introduce a new approach— intensification of leadership - that affects teaching and learning and will change the cultural conditions of a school. Briefly, it is a way to increase the number of people engaged in leadership roles and the scope of the school's work as it relates to student outcomes. In concrete, there are three conditions for constructing school cultures captured by the PCOLT strategy: professional community, organizational learning, and trust (8–9).

The concept of a professional community is based on the authors' belief that deep-seated changes in the culture of schools are unlikely to occur without action to create more fundamental bonds within the community. Accordingly, they argue that strong school cultures are based on shared norms and values, reflective dialogue, public practice, and collaboration. Organizational learning is the concept that collective engagement with new ideas will generate enhanced classroom practices and a deeper understanding of how organizational improvement occurs. Trust is an essential factor for a great and strong school culture as it is "the glue that holds social networks and relationships together" (Kruse and Louise, 2009: 9). A school organization with high levels of trust internally will observe effective communication, openness, and positive attitudes towards seeing disagreement and conflict as opportunities for problem-solving.

Preble and Gordon's research (2011) suggests that a good school climate/culture should contain respectful teaching and school practices. The quality of school climate/culture is closely related to quality of the relationships within a school, its overarching vision, goals and mission, and the support systems for students, teachers, and parents that enable the school to achieve its mission. Meanwhile the roles available to and played by students, teachers and school leaders and the opportunities for active involvement and meaningful engagement as learners, leaders and citizens within the school and community, and extent to which there is respect, tolerance, fairness, equity, and social justice are other significant influence on school culture (2011: 2).

Despite diverse perspectives and approaches within recent research, an agreement can be reached on the importance and essence of school culture.

Necessary components consist of clear vision, goals and objectives, core values, cooperation and collaboration, commitment, team orientation, and professional development. This study shares the view that school culture is highly complex and is where a web of traditions and rituals are established overtime with the participation of teachers, students, parents, and administrators (Schein, 1985, Deal and Peterson, 1990, 1999, 2009).

2.3 School Culture in China

In the past decade, school culture has gained increased attention from scholars in China. Research focuses vary from philosophy, theory, and introduction of Western experiences to the Chinese context to analysis and critical reflection on the current situation and practices in China. In general there is a widespread consensus that schools in China need significant renovating. In a broad sense, scholars (Shi, 2005; Gu, 2006; Deng and Xi, 2007; Yang, 2009; Wang, 2012; Wu, 2013) agree that school culture encompasses spiritual culture, institutional culture, and the campus material culture, with values as the core of school culture. The cultural construction of schools is the unification of inheritance and innovation and is the result of the educational practices of leaders, teachers, and students over time.

Zhu et al. (2011) summarize three levels of research on school culture in China: the ideological level (value, ways of thinking, beliefs, etc.), the (assumed) behavioral level (behavior code, ethics, customs, public opinion, etc.), and the material level (clothing, architecture, physical environment, etc.). Focusing on the material level tends to be unique to the studies conducted in China when compared to international literature.

Despite the substantial growth in literature on school culture in China, this study finds that the majority of Chinese literature is primarily focused on the following aspects: Firstly, the introduction of the concept and a presentation of the research field based on experiences from Western countries; Secondly, the argument that school culture should be given more attention in the Chinese context; Thirdly, the recognition of the need to enhance school culture at practical levels from a policy perspective; Fourthly, a descriptive introduction to some cases of school culture construction in China.

Recent literature also voices concerns and critiques of school culture development in China. For example, more attention is given to campus construction than spirit construction, symbolic construction emphasizes logos over actions to create changes in practices, and there is a general lack of regulation at the policy making level (Deng and Xi, 2007; Yang, 2009; Wang, 2012).

In general, current research and practice in the construction of school culture in China remains unsystematic and needs to be developed. Scholars (Shi, 2005; Gu, 2006; Deng and Xi, 2007) cite the following aspects as reasons for these problems. Firstly, understandings of school culture and its nature are varied among researchers and practitioners. Secondly, usage of an evidence-based approach to constructing school culture is lacking. Thirdly, there is lack of attention to teachers' understandings and assumptions of, as well as attitudes towards school culture. Finally, developing appropriate leadership in terms of constructing school culture must still be developed.

The literature in general makes a positive contribution to the development of schools and the research field of school culture by taking theoretical inspiration and experiences from the West. However, this study argues that there has been a lack of empirical studies that focus on the diagnosis and change process in the field of school culture. As attention has been called to an evidence-based approach and research-based method to further improve the development of school culture work (Cui and Zhou, 2007; Wu, 2013).

Proposals arguing for schools to be refurbished in order to give more empowerment to leadership, administration and teachers is also frequently invoked (Shi, 2005; Gu, 2006; Deng and Xi, 2007; Yang, 2009; Wang, 2012; Wu, 2013). Therefore, this study aims to analyze and understand school culture in terms of its characteristics and dimensions. It also attempts to investigate teachers' and principals' perceptions and understandings of school culture. The results of this pilot study will attempt gain a clear view of the current status of school culture and will contribute to the further development of strategies for school culture work in China. This study is also expected to contribute to the field of school culture development through its explanation of employing an empirical evidence-based approach.

2.4 Focuses of this Study

2.4.1 Perspectives from Teachers and Principals

In school culture studies, major participants and subjects are often students, principals, teachers, and parents. In this volume, we mainly focus on the perceptions of teachers and principals because from an organizational perspective, staff's well-being and commitment to an organization are important factors that influence organizational effectiveness and improvement (Deal and Peterson, 2009).

Teachers are playing an increasingly active role in school development in terms of their growing involvement in curriculum development and other decision-making bodies. However, in the past decade, research has shown that an increasing amount of pressure has been placed on school teachers who are expected to expand curriculum on reduced resources. Beaudoin and Taylor (2004) identify four main pressures that place significant stress on school teachers such as sacrificing personal time, being largely responsible for many issues, always being in control, and being perfect role models for their students. These stresses result in: teachers' own health, family, and personal space needs being sacrificed; feeling drained and facing professional burnout in some cases; feeling guilt and frustration; and having reduced self-confidence in their profession. These pressures and negative effects on teachers reduce their ability to access possible solutions to challenges and find creative and innovative ways of doing things; in turn, this reduced innovation negatively impacts student learning and school success. Helping school teachers cope with the issues of negativity, isolation and condemnation calls for research attention and proposing practical solutions.

School principals tend to believe that they understand teachers' experiences because they have been teachers themselves (Beaudoin and Taylor, 2004). However, this awareness of teacher needs has been disproven. Principals often prefer to remember what happened in the classroom and tend to forget the nuances experiences when their career developed progressively (Beaudoin and Taylor, 2004). As principals, they tend to see things differently than teachers. From this perspective, an assumption can be made that teachers and principals have differing views on what school culture is and how the environment they are situated in should be constructed. This assumed difference could possibly create barriers to the sustainable development of schools and student learning.

Nowadays, school leaders are not only expected to focus on students' performance and curriculum development, but are also expected to pay attention to a much broader scope of organizations that connect the school with foundations, business groups, social service providers, and government agencies (Kruse and Louise, 2009). An effective school leader is expected to lead change of the school culture in order to maximize teaching and learning by providing a vision, providing encouragement and recognition, obtaining resources, adapting standard operating procedures, monitoring school improvement efforts, and handling disturbances (Heller and Firestone, 1995).

However, managing a school's culture is not only dependent on the authority of the principal but is also determined by collective efforts with teachers, students, and parents. Principals must understand and support what teachers do in the classrooms in order to help create the conditions that allow them to be more effective.

Deal and Peterson (2009) examine the way leaders shape culture to create a cohesive, meaningful, nurturing, and social milieu for teachers to teach and students to learn. Leadership in robust cultures is dispersed among teachers, administrators, parents, and students. Together they read, shape and continuously renovate the culture of their school (Deal and Peterson, 2009).

One of the most significant roles for leaders is the creation, encouragement, and refinement of the symbols and symbolic activity that confer meaning (Deal and Peterson, 2009:15). Leadership in schools must balance a dueling emphasis on maintaining stability while creating change. These roles and responsibilities provide space for principals to be innovative in schools and to create new and supportive school cultures; nevertheless, these opposing responsibilities also give rise to pressures and stresses of the school leadership (Beaudoin and Taylor, 2004).

In summary, this study holds the view that it is crucial that teachers and principals share common views towards what a good school culture should be, and how they can build up a good school culture together given the specific conditions and spheres they are in. Based on this viewpoint, the first step of the research project will investigate how principals and teachers understand school culture, and in particular, how school culture is perceived at their school.

2.4.2 An Orgnizational Perspective on and Approach to School Culture

The study of school culture has often been related to organizational theories, since school culture is often considered a branch of organizational culture in an educational context (Schoen and Teddlie, 2008). One of the primary beliefs in the field is that culture plays an important role in organizational behavior. As Deal and Peterson (2009:1) state, "the culture of an enterprise plays a dominant role in exemplary performance." From this perspective, some studies have been conducted to investigate internal cultures of specific organizations such as schools (Deal and Peterson, 1999; 2009). In academic debates over school culture change, one influential argument is that schools should be revamped to more closely resemble businesses. Deal and Peterson state that (2009:16), "top businesses have developed a shared culture with a deep set of values." A successful firm's culture drives meaning, passion, and purpose into the enterprise, and employees of such organizations devote their hearts and souls to the institutional practices. It is argued that the same passion and purpose should be found in educational institutions such as schools. Therefore, the study takes an organizational perspective and diagnoses school culture via an organizational study approach from the perspectives of teachers and principals.

Organizational culture is considered to be one of the most important factors in carrying out daily business, company unification, and organizational performance, among other functions, in all kinds of organizations. A previous study has suggested that there is no ideal instrument for cultural exploration, and the fitness of any measurement depends on the purpose for which it is to be applied and the context within which it is to be used (Jung et al., 2009).

Studies of organizational culture, and, in particular, of school culture, are often dominated by qualitative approaches. Many researchers believe that cultures cannot be measured and compared (Schoen and Teddlie, 2008). Qualitative research has also supported the notion that school culture may be related to outcomes for a school. For example, in Van der Westhuizen's study (2008), formal and informal observation and interviews were employed in a boys' boarding house in order to find the link between organizational culture and student discipline. The authors concluded that there is positive relationship between organizational culture and student discipline, and effective student discipline and effective organizational culture can be dependent upon each other.

After years of debate, some researchers have proposed that organizational culture consists of different levels, and that different research methods are required for the different levels. This perception argued that culture could be measured at certain or varying levels. Schein (1992) described the three levels of organizational culture as artifacts, espoused beliefs, and basic assumptions. The first level of Schein's theoretical structure is basic assumptions, which are unconscious, taken-for-granted beliefs, perceptions, thoughts, and feelings that are shared by members of the organization. This level should be investigated using observation and loosely or unstructured interviews. Likewise, the third level, artifacts, meaning visible organizational structures and processes in the organization, should also be researched through observation and interviews. The second or intermediate level, however, espoused beliefs, referring to strategies, goals, and philosophies of the organization, should be investigated using surveys and structured interviews.

Questionnaires and various other research instruments have been developed that are in line with Schein's work. For instance, Cooke and associates (1988) designed the Organizational Culture Inventory to measure culture at the behavioral-norms level. The Inventory identifies three types of cultural styles: constructive, passive/defensive, and aggressive/defensive. After investigating 20 Dutch and Danish firms, Hofstede and associates (1993) defined six dimensions of organizational culture. These dimensions are: process/results orientation, employee/job orientation, parochial/professional orientation, open/closed system, loose/tight, and normative/pragmatic. This questionnaire can be used to measure national differences in organizations.

Cameron and Quinn (1999) developed an organizational culture framework based on the "Competing Values Framework" theoretical model. This culture framework distinguishes organizations based on two dimensions, namely 1) internal vs. external focus and 2) flexibility and individuality vs. stability and control. In addition, the "Organizational Culture Assessment Instrument" (OCAI) was invented based on this organizational culture framework. By investigating organizations' core values, assumptions, interpretations, and approaches, this instrument can identify four dominant culture types: clan, adhocracy, market, and hierarchy.

Goffee and Jones (1998) focused their organizational culture work on the willingness to share knowledge in an organization. Their organizational culture model is based simply on the sociability and solidarity perspectives. Sociability refers to the reciprocal relationships between people, and solidarity refers to relationships between individuals and the wider organization. By combining these two dimensions, four cultural orientations can be identified: communal culture, networked culture, fragmented culture, and mercenary culture.

The "Organizational Health Inventory" (OHI) is theory driven. It is comprised of seven dimensions used to evaluate organizational function on three levels of responsibility and control: technical, managerial, and institutional. Each level is measured by instrumental activities, expressive activities, or both types of activities. Seven dimensions generate one second-order factor called health. Healthy schools can function well (Hoy and Feldman, 1987).

In addition to instruments used to identify specific organizational type and measure differences, some questionnaires have been developed for the purpose of employee selection and socialization. For instance, Chatman's (1991) work was used to investigate how closely employees and organizations can match and to predict employee satisfaction. There are eight dimensions in this approach: innovation, attention to detail, outcome orientation, aggressiveness, supportiveness, and emphasis on rewards, team orientation, and decisiveness. Every questionnaire was developed in a specific context for specific purposes. As indicated in a recent review, each approach offers different insights into exploring organizational culture (Jung et al., 2009).

Schools as a special kind of organization also have their own climates or cultures. Halpin (1966) realized that schools have quite different atmospheres than other kinds of organizations. He conceptualized organizational climate or culture and created the "Organizational Climate Descriptive Questionnaire" (OCDQ). It is empirically driven. The first edition consisted of eight dimensions. Four of them are about the staff and faculty members as one group. The other four are about teacher-principal interaction. Through analysis, six basic school climates were identified. Those six climates were positioned from open to closed. The OCDQ was used widely. In order to respond to criticisms of the original version, a new version for elementary school (Hoy and Clover, 1986) and one for secondary school (Kottkamp, Muhern, and Hoy, 1987) were developed.

Denison's organizational culture model and the "Denison Organizational Culture Survey" are frequently used in the field of organizational studies worldwide. Denison's organizational culture model focuses on organizational culture and organizational effectiveness (Denison et al., 2006). This approach directly emphasizes those aspects of organizational culture that appear to influence organizational effectiveness. The model is based on four cultural traits: involvement, consistency, adaptability, and mission. Each of these traits consists of three component indexes. Involvement is comprised of empowerment, team orientation, and capability development. Consistency includes core values, agreement, coordination, and integration. Adaptability consists of creating change, customer focus, and organizational learning. Mission consists of strategic direction and intent, goals and objectives, and vision. The Denison Organizational Culture Survey was established and implemented on the basis of these traits.

One of the major reasons that the Denison Organizational Culture Survey has been so often employed worldwide is that many studies have shown that organizational culture is related to organizational effectiveness, encompassing such factors as employee satisfaction, customer satisfaction, profits, etc. (Gillespie, 2008). Organizational culture can be a predictor of a company's achievement. Therefore, improving company culture may be an effective way for companies to increase profits, efficiency, and opportunities for sustainable development.

Therefore, in most cases, the Denison Organizational Culture Survey is used as a diagnostic instrument to help an organization understand the advantages and disadvantages of the company's culture. After evaluation, the organization provides improvement-oriented training activities that focus specifically on the indexes that received lower values in the diagnostic process. Many organizations have reported benefits from using this instrument.

In summary, the Denison Organizational Culture Survey is employed as an instrument in this study to investigate how teachers and principals perceive school culture development at their schools for two reasons. Firstly, the survey and its theoretical background are in line with the process approach to the conceptual understanding in this study, which lays emphasis on the complex nature of culture and school culture. Secondly, the survey offers opportunities to understand and diagnose school culture and allows for the possibility of recommendations with regard to further improvement. This aspect of the survey can be seen as an advantage that fits the research purpose of the present study. The choice of method is, at the same time, challenged by the fact that the Denison Organizational Culture Survey has not been employed in an educational context at the school level, nor, in particular, in a Chinese educational context. We further introduce the method and research context of this study in the next chapter.

3

Method

This chapter gives an overview of the research methods in this study — design and contents of the survey, data generation and analysis, and limitation of the methods. This introduction is followed by descriptive data from the respondents, concerning the overall number of participants, gender, educational level, job duty, job title, years in teaching, and school location. This chapter lays a methodological and empirical foundation for the following chapters of the book.

3.1 Research Methods

The Survey

As introduced in chapter 2, Denison's organizational culture model consists of four traits: involvement, consistency, adaptability, and mission. Each of these traits consists of three component indexes, and each of these indexes is measured by five items. Thus, there are 60 items in the Denison Organizational Culture Survey. Denison's model and survey structure have been confirmed by confirmatory factor analysis by using 35,474 participants in 160 organizations all over the world. This survey has also shown high reliability indicated by high alpha coefficients. The validity of this survey is also good, as indicated by six measures of organizational performance. Furthermore, within the academic community, this survey is considered a to be a valued measurement in the field (Denison et al., 2013).

The Denison Organizational Culture Survey has been used in a variety of sectors, such as the financial sector, health care, materials science, energy, technology, etc. It is popular in many countries, including Australia, France, Japan, Great Britain, Switzerland, and the United States. Some researchers have translated and adapted the Denison Organizational Culture Survey into their own languages and tested the reliability and validity of the translated surveys. For example, the Icelandic version kept all 60 original items (Gudlaugsson,

2013). The psychometric properties of the Icelandic version were tested based on the responses of 1,132 participants in 13 Icelandic firms. The four traits and twelve indexes showed acceptable internal consistency. In the factor analysis, the mean of five items was taken as the score for the corresponding index to lower the number of parameter estimation. Confirmative factor analysis of these twelve indexes showed acceptable goodness of fit of the four-trait model as compared to the original version. Fey and Denison translated the survey items into Russian and applied the Denison Organizational Culture Survey with participants from foreign firms in Russia (Fey and Denison, 2003). The Russian translation includes four traits, each with three indexes, as in the original version. However, each index in this Russian version was measured by three items, rather than the five items of the original. Factor analysis confirmed the four-trait model. Cronbach alphas showed good internal reliability for all four traits. All four traits were significantly related to organizational effectiveness. Thus, the Russian version showed good reliability and validity.

In this pilot study, the Denison Organizational Culture Survey was employed as the instrument of investigation (with the permission of Denison Consulting). Slight revisions were made in the translation process to better fit the context of a study of school culture in China from the perspective of the education field and to better fit the Chinese language (which was used to conduct the survey). Two versions (with minor differences) were used in the present study, one for teachers (see Appendix 8) and the other for principals (see Appendix 9). They differed only in item ordering.

As in the original version, there are four traits in the Chinese school version, namely involvement, consistency, adaptability, and mission. Each trait is comprised of three indexes. Unlike the original version, each index consists of four items. There are 48 items altogether in the questionnaire (see Appendix 10) These items are scored on a five-point Likert-type scale. Participants were asked to evaluate to what extent they agreed with a statement on a scale of 1 to 5, with 1 being "totally disagree" and 5 being "totally agree." A sample item is: "Most teachers at our school are deeply involved in their work." There are eight reversed items. An example of a reversed item is: "Problems often arise because some teachers do not have the skills necessary to do the job". These reversed items were recoded afterwards.

The first stage of data collection in this pilot study was finalized by autumn 2013. A total of 2066 participants took part in the survey and answered the questionnaire. Of these participants, 1,992 people (96.42%) returned valid questionnaires that were included in the final data analysis. Data on respondents' gender, educational level, job duty, job title, years in teaching, and affiliation were also collected.

The internal reliability of this questionnaire and of each trait was satisfactory.

For the trait of involvement, Cronbach's α is 0.926. Involvement consists of *empowerment, team orientation,* and *capability development.* The reliability of each index was also satisfactory. Detailed information on each index and item is presented in Table 3.1.

Table 3.1 Internal reliability of involvement

Index Cronbach's α	Item	Cronbach's α if the item is deleted
Empowerment 0.848	Most teachers are highly involved in their work.	0.830
	Everyone believes that he or she can have a positive impact.	0.812
	Planning is ongoing and involves everyone in the process to some degree.	0.794
	Information is widely shared so that everyone can get the information he or she needs when it's needed.	0.785
Team Orientation 0.884	Cooperation across different parts of the school is actively encouraged.	0.869
	Sufficient cooperation among teachers is taking place in ethical education, teaching, and other areas.	0.840
	Teamwork, is encouraged to get school work done, rather than hierarchy.	0.852
	Each teacher can see the relationship between his or her job and the goal of the school.	0.842
Capability Development 0.637	Authority is delegated so that teachers can act on their own.	0.639
	The professional skills of teachers are constantly improving.	0.445
	There is continuous financial investment from the school to improve the professional skills of the teachers.	0.454
	Problems often arise because some teachers do not have the skills necessary to do the job. (Reversed Scale)	0.704

For trait of Consistency, Cronbach's α is 0.918. Consistency is consisted of *core values, agreement, coordination and integration.* The reliability of each index is also satisfactory. The detailed information on each index and each item was presented in Table 3.2.

For trait of Adaptability, Cronbach's α is 0.867. Adaptability is consisted of *creating change, customer focus,* and *organizational learning.* The reliabilities of creating change and organizational learning are satisfactory. The reliability of the customer is low (Cronbach's α = 0.523). The reliability would be acceptable if the reversed item "The interests of the parents often get ignored in our decisions" was deleted. The detailed information on each index and each item was presented in Table 3.3.

For trait of Mission, Cronbach's α is 0.926. Mission is consisted of *strategic direction and intent, goals and objectives,* and *vision.* The reliability of each index is also satisfactory. The detailed information on each index and each item was presented in Table 3.4.

Data Analysis

The software SPSS15.0 was used for data analysis. Mean (M) and Standard Deviation (SD) were reported in descriptive results, as well as the number (N) of participants. For each item, t-test was used for the comparison between

Table 3.2 Internal reliability of consistency

Index Cronbach's α	Item	Cronbach's α if the item deleted
Core Values 0.903	I am aware of the principal's educational philosophy and can adjust my own accordingly.	0.868
	I understand the school's development goals.	0.859
	I understand the school motto's cultural meaning.	0.861
	My mode of dress and behavior is consistent with school culture.	0.907
Agreement 0.759	When disagreements occur among teachers, we work hard to achieve "win-win" solutions.	0.664
	We always reach consensus, even on difficult issues.	0.646
	There is a clear agreement about the right way and the wrong way to do things.	0.639
	We often have trouble reaching agreement on key issues. (Reversed Scale)	0.849
Coordination and Integration 0.752	People from different parts of the school share a common perspective.	0.643
	It is easy to coordinate working across different parts of the school.	0.637
	It is difficult for teachers to cooperate in our school. (Reversed Scale)	0.855
	There is good alignment of goals across all levels.	0.618

Table 3.3 Internal reliability of adaptability

Index Cronbach's α	Item	Cronbach's α if the item deleted
Creating Change 0.737	The mode of both teaching and management is very flexible.	0.602
	The school has an effective strategy for competing with peer schools.	0.621
	New and improved ways to do work are continually adopted.	0.590
	Attempts to create change usually meet with resistance. (Reversed Scale)	0.870
Customer Focus 0.523	Parent comments and recommendations often lead to changes.	0.558
	All teachers have a deep understanding of parents want and need.	0.264
	The interests of the parent often get ignored in our decisions. (Reversed Scale)	0.616
	We encourage teachers to have direct contact with parents.	0.300
Organizational Learning 0.712	We view failure as an opportunity for learning and improvement.	0.595
	Innovation and risk taking are encouraged and rewarded.	0.593
	Lots of things "fall between the cracks". (Reversed Scale)	0.777
	Learning is an important objective in our day-to-day work.	0.627

two groups. Analysis of variance (ANOVA) was used for the comparison among three or four groups. When significant difference among these three or four groups was found, the least significant difference (LSD) post-hoc test was used. $P < .05$ was considered statistically significant. The detailed participants' distribution on gender, educational level, job duty, job title, years in teaching, and school location are described in the following sections.

Limitation of the Method

The design of this research is limited in a few ways for certain practical reasons. Firstly, the questionnaire employed in the present research was adapted from organizational field. This may be questioned by its validity in school culture studies. Secondly, several reversed items were used. Participants may find it is difficult to switch between reversed and right counting items. Thirdly, the participants of this study were recruited from four districts in Beijing. The

Table 3.4 Internal reliability of mission

Index Cronbach's α	Item	Cronbach's α if the item deleted
Strategic Direction & Intent 0.702	Peer schools in the district wish to imitate our development strategy.	0.675
	All school work is conducted with the guidance of school development strategies.	0.528
	There is a clear strategy for the future.	0.537
	Our strategic direction is unclear to me. (Reversed Scale)	0.799
Goals & Objectives 0.893	Leaders set goals that are ambitious, but realistic.	0.850
	The leadership has "gone on record" about the objectives we are trying to meet.	0.855
	Leaders continuously track our progress against our stated goals.	0.865
	I will devote myself to fulfilling the school's development goals.	0.875
Vision 0.781	Teachers have a shared vision of what the school will be like in the future.	0.693
	Leaders have a long-term viewpoint.	0.671
	Short-term thinking often compromises our long-term vision. (Reversed Scale)	0.863
	Our vision creates excitement and motivation for teachers.	0.679

process did not follow stratification sampling. The results of this study cannot be generalized to other schools, such as schools in rural area.

3.2 Respondents' Information

Respondents' backgrounds

Respondents in the investigation came from 37 primary and middle schools in Beijing, China. The number of participants from these schools varied between 22 and 126. The choice of participant school (instead of choosing random schools) was based on a pragmatic approach to enable easy access to data. Although this has the potential limitation of shadowing the research results, it is an effective way to increase the reliability of this research.

Gender

Among the 1992 valid respondent questionnaires, there were 355 males, 1621 females and 16 without gender identification. Gender distribution in the study was not equally weighted. There are more females than males among the participants (see Table 3.5).

Educational level

These 1992 participants were of four different educational levels, namely Postgraduate degree, Bachelor's degree, Diploma, and Other. The group of Bachelor's degree (1534 people) has a considerably higher number of participants than the other groups (see Table 3.6).

Job duty

Participants' job duty was also taken into consideration in this study. 1883 out of the 1992 valid questionnaires stated their job duty. The group of teacher had more participants (1719 people) than groups of middle leadership and principals (see Table 3.7).

Job title

Three hierarchic terminologies are often used for job titles in the Chinese academic world; Junior, Middle, and Senior in ascending order. Among the 1992 valid respondent questionnaires, all three titles were represented: 424 participants with a Junior title, 470 with a Middle title, and 1001 with a Senior title (see Table 3.8). There were more participants in Senior group than in Junior and Middle group.

Teaching year

Participants were of different teaching years. They were divided into three groups, those with less than 6 years of teaching experience, 6–15 years of experience, and more than 15 years of teaching experience. The participants' number of years in teaching group was not equally distributed (see Table 3.9). There were fewer participants in the group with less than six years of experience than the group with 6–15 years of experience and the group with more than 15 years.

Table 3.5 Gender distribution

	Gender	N	Percent	Valid Percent	χ^2	P
Valid	Male	355	17.8	18.0	811.11	<0.001
	Female	1621	81.4	82.0		
	Total	1976	99.2	100.0		
Missing		16	0.8			
Total		1992	100.0			

Table 3.6 Participant distribution byeducational level

	Education	N	Percent	Valid Percent	χ^2	P
Valid	Postgraduate	74	3.71	3.83	3130.57	<0.001
	Bachelor's degree	1534	77.01	79.48		
	Diploma	286	14.36	14.82		
	Other	36	1.81	1.87		
	Total	1930	96.89	100.00		
Missing		62	3.11			
Total		1992	100.00			

Table 3.7 Participant distribution by job duty

	Job duty	N	Percent	Valid Percent	χ^2	P
Valid	Teacher	1719	86.30	91.29	2852.45	<0.001
	Middle leadership	126	6.33	6.69		
	Principal	38	1.91	2.02		
	Total	1883	94.53	100.00		
Missing		109	5.47			
Total		1992	100.00			

Table 3.8 Participant distribution by job title

	Jobtitle	N	Percent	Valid Percent	χ^2	P
Valid	Junior	424	21.29	22.37	325.60	<0.001
	Middle	470	23.59	24.80		
	Senior	1001	50.25	52.82		
	Total	1895	95.13	100.00		
Missing		97	4.87			
Total		1992	100.00			

School Geographical Location

Participants were from 37 primary and middle schools in four different geographical districts (District A, B, C, and D) within Beijing Municipality. According to one of the ways used to categorize districts by the Beijing Educational Commission, District A is considered a developing district, meaning that it is an economically less developed area within Beijing Municipality and that the overall educational vision of this district is to provide basic education to all students. Historically District A has devoted fewer resources to education in general and innovation projects in particular compared with other districts. Accordingly, school principals hired in this district are less qualified in terms of educational background and capabilities. The district is populated with migrant families, with 50 percent of the students' parents having low levels of education and holding low-paying temporary jobs, such as those at

Table 3.9 Participant distribution by years in teaching

	Teaching Year	N	Percent	Valid Percent	χ^2	P
Valid	<6	382	19.18	19.53	234.00	<0.001
	6–15	640	32.13	32.72		
	>15	934	46.89	47.75		
	Total	1956	98.19	100.00		
Missing		36	1.81			
Total		1992	100.00			

Table 3.10 Participant distribution by school location

Location	N	Percent	χ^2	P
District A	361	18.12	833.90	<0.001
District B	1011	50.75		
District C	133	6.68		
District D	487	24.45		
Total	1992	100.00		

construction sites and in service industries. No school culture development project had been initiated at the time this pilot study was conducted.

Districts B, C, and D are comparatively developed districts, meaning that they have more financial resources devoted to educational development than is the case in District A. The three districts all to different degrees have a history of successfully implementing educational innovations and of higher academic achievement compared with District A. Districts B, C, and D have all had a few ongoing projects related to school culture development in the past few years. There are no significant differences among these three districts with regard to their economic situations or achievement in school culture development projects. There are, in general, more schools, classes, teachers, staff, and students in District C than in Districts A, B, and D.

There were significant differences in the number of participants from each of these four districts (see Table 3.10). There were more participants from District B than from District A, C, or D.

Summary

This chapter introduces the background of the research methods for this book, which includes the design and contents of the survey, data generation and analysis, and the limitation of the methods. It also presents the descriptive data of the respondents' information concerning the overall number of participants, their gender, educational level, job duty, job title, years in teaching, and school location. This chapter functions as a methodological and empirical basis for the following chapters of the book.

4

Involvement

This chapter explores school staff members' perceptions of one of the school culture traits: *involvement*. Involvement consists of three indexes, namely *empowerment, team orientation,* and *capability development*. Twelve survey items were used as a way to evaluate these indexes. The findings are presented and analyzed in three sections. Section one reports how participants responded to questions about teachers' involvement in their work, as well as teachers' beliefs regarding their impact on the school's development, opportunities to participate in the school development planning, and their possibility to access the school's information. Section two presents participants' perceptions of cooperation across different parts of the school, cooperation in ethical education, teaching, and other educational areas, the encouragement of teamwork, and teachers' understanding of the relationship between their jobs and the school's development. Section three shows how participants responded to questions about teachers' authority, improvements in teachers' professional skills, financial investments made by the school in improving teachers' professional skills, and the problems that can arise when teachers do not have the skills necessary to do their jobs. Each section details the overall findings for the given items as well as the effects of participants' gender, educational level, job duty, job title, years in teaching, and school location on the findings.

4.1 Empowerment

In order to understand participants' perceptions of the empowerment of their school culture, four items - "Most teachers are highly involved in their work;" "Everyone believes that he or she can have a positive impact;" "Planning is ongoing and involves everyone in the process to some degree;" "Information is widely shared so that everyone can get the information he or she needs when it's needed" - were used that required five-point Likert-type responses.

Overall Findings

A summary of the descriptive data of the participants' responses to these items is shown in Table 4.1. In general, participants reported higher values than the mean on all items, especially on the first one. Participants held a positive perception of their school's level of empowerment.

Influence of Gender

Both male and female participants took a favorable view of their school's degree of empowerment (see Table 4.2). In general, females assessed teachers' involvement in their work and teachers' beliefs in their impact on the school's development more positively than males. Males and females had similar perceptions of the opportunity to participate in the school development planning and the availability to access school information.

Table 4.1 Participants' perceptions of empowerment

Item	N	M	SD
Most teachers are highly involved in their work.	1986	4.50	0.79
Everyone believes that he or she can have a positive impact.	1973	4.31	0.88
Planning is ongoing and involves everyone in the process to some degree.	1973	4.03	1.08
Information is widely shared so that everyone can get the information he or she needs when it's needed.	1982	4.21	1.01

Table 4.2 Participants' perceptions of empowerment by gender

Item	Gender	N	M	SD	t	P
Most teachers are highly involved in their work.	Male	353	4.36	0.88	–3.78	<0.001
	Female	1617	4.54	0.76		
Everyone believes that he or she can have a positive impact.	Male	352	4.22	0.94	–1.98	0.048
	Female	1605	4.33	0.86		
Planning is ongoing and involves everyone in the process to some degree.	Male	352	4.08	1.03	0.83	0.407
	Female	1605	4.03	1.09		
Information is widely shared so that everyone can get the information he or she needs when it'sneeded.	Male	352	4.14	1.08	–1.58	0.113
	Female	1615	4.23	0.99		

Influence of Educational Level

All four educational level groups had positive perceptions of their school's empowerment (see Table 4.3). There was no significant difference among the four educational level groups on any of the items.

The *P*-values of post-hoc tests between all possible combinations of educational level groups are shown in Table 4.4.

Influence of Job Duty

All three job duty groups held positive opinions of their school's performance in the area of empowerment (see Table 4.5). Teachers, middle leadership, and principals had similar perceptions of teachers' involvement in their work and the availability of and access to school information. However, the groups differed in their perceptions of teachers' impacts on the school's development and teachers' opportunity to participate in school development planning.

The *P*-values of post-hoc tests between all possible combinations of job duty groups are shown in Table 4.6. Teachers and middle leadership reported higher values than principals on teachers' beliefs on their impact on the school's development. Middle leadership reported higher values than teachers on teachers' opportunity to participate in the school development planning.

Table 4.3 Participants' perceptions of empowerment by educational level

Item	Education	N	M	SD	F	P
Most teachers in our school are deeply involved in their work.	Postgraduate degree	74	4.41	0.74	2.35	0.070
	Bachelor's degree	1533	4.49	0.80		
	Diploma	281	4.61	0.68		
	Other	36	4.53	0.77		
Everyone believes that he or she can have a positive impact.	Postgraduate degree	74	4.27	0.82	0.13	0.943
	Bachelor's degree	1524	4.31	0.87		
	Diploma	279	4.30	0.89		
	Other	34	4.24	0.92		
Planning is ongoing and involves everyone in the process to some degree.	Postgraduate degree	74	4.08	0.96	0.17	0.918
	Bachelor's degree	1522	4.03	1.08		
	Diploma	279	4.06	1.07		
	Other	36	4.11	1.06		
Information is widely shared so that everyone can get the information he or she needs when it's needed.	Postgraduate degree	73	4.22	0.95	0.11	0.954
	Bachelor's degree	1529	4.22	1.00		
	Diploma	283	4.20	1.03		
	Other	36	4.14	1.10		

Table 4.4 Participants' perceptions of empowerment by educational level: Post-hoc test

Item	Postgraduate vs. Bachelor	Postgraduate vs. Diploma	Postgraduate vs. Other	Bachelor vs. Diploma	Bachelor vs. Other	Diploma vs. Other
Most teachers are highly involved in their work.	0.391	0.047	0.442	0.015	0.748	0.560
Everyone believes that he or she can have a positive impact.	0.710	0.812	0.847	0.839	0.627	0.696
Planning is ongoing and involves everyone in the process to some degree.	0.687	0.886	0.891	0.654	0.653	0.792
Information is widely shared so that everyone can get the information he or she needs when it's needed.	0.979	0.893	0.694	0.747	0.622	0.725

Table 4.5 Participants' perceptions of empowerment by job duty

Item	Job duty	N	M	SD	F	P
Most teachers are highly involved in their work.	Teacher	1715	4.52	0.79	0.53	0.589
	Middle leadership	125	4.50	0.64		
	Principal	38	4.39	0.64		
Everyone believes that he or she can have a positive impact.	Teacher	1705	4.33	0.87	6.05	0.002
	Middle leadership	123	4.42	0.81		
	Principal	38	3.87	0.81		
Planning is ongoing and involves everyone in the process to some degree.	Teacher	1705	4.02	1.09	5.98	0.003
	Middle leadership	123	4.37	0.85		
	Principal	38	4.13	0.78		
Information is widely shared so that everyone can get the information he or she needs when it's needed.	Teacher	1715	4.22	1.01	2.32	0.099
	Middle leadership	123	4.40	0.79		
	Principal	38	4.37	0.79		

Table 4.6 Participants' opinions on empowerment by job duty: Post-hoc test

Item	Teacher vs. Middle Leadership	Teacher vs. Principal	Middle leadership vs. Principal
Most teachers are highly involved in their work.	0.737	0.325	0.481
Everyone believes that he or she can have a positive impact.	0.233	0.001	0.001
Planning is ongoing and involves everyone in the process to some degree.	0.001	0.535	0.240
Information is widely shared so that everyone can get the information he or she needs when it's needed.	0.049	0.348	0.871

Influence of Job Title

All three job title groups felt positively about their school's empowerment (see Table 4.7), though there were significant differences among the three groups on all four items.

The P-values of post-hoc tests between all possible combinations of job title groups are shown in Table 4.8. On the teachers' involvement in their work, Junior and Senior groups reported higher levels of involvement than the Middle group. With regard to teachers' beliefs on their impact on the school's

Table 4.7 Participants' perceptions of empowerment by job title

Item	Job title	N	M	SD	F	P
Most teachers are highly involved in their work.	Junior	423	4.55	0.73	10.94	<0.001
	Middle	467	4.36	0.86		
	Senior	999	4.55	0.77		
Everyone believes that he or she can have a positive impact.	Junior	423	4.30	0.87	8.50	<0.001
	Middle	459	4.18	0.88		
	Senior	996	4.38	0.87		
Planning is ongoing and involves everyone in the process to some degree.	Junior	423	4.04	1.05	16.50	<0.001
	Middle	462	3.82	1.09		
	Senior	991	4.17	1.06		
Information is widely shared so that everyone can get the information he or she needs when it's needed.	Junior	422	4.20	1.02	15.01	<0.001
	Middle	466	4.03	1.04		
	Senior	998	4.33	0.96		

Table 4.8 Participants' perceptions of empowerment by job title: Post-hoc test

Item	Junior vs. Middle	Junior vs. Senior	Middle vs. Senior
Most teachers are highly involved in their work.	<0.001	0.893	<0.001
Everyone believes that he or she can have a positive impact.	0.051	0.087	<0.001
Planning is ongoing and involves everyone in the process to some degree.	0.003	0.034	<0.001
Information is widely shared so that everyone can get the information he or she needs when it's needed.	0.010	0.024	<0.001

development, the Middle group reported lower values than the Senior group. Concerning teachers' opportunities to participate in school development planning, the Senior group reported more opportunities for participation than Junior group, who in turn reported having more opportunities than the Middle group. Regarding the availability of and access to school information, the Senior and Junior groups reported higher values than the Middle group.

Influence of Number of Years in Teaching

All three of the teaching experience groups, those with less than 6 years of teaching experience, 6–15 years of experience, and more than 15 years of teaching experience, held positive opinions of their school's empowerment (see Table 4.9). There were no significant differences among the three groups on any of the items.

The P-values of post-hoc tests between the various groups according to the number of years in teaching are shown in Table 4.10.

Influence of the School's Geographical Location

All four school geographical location groups had positive views of their school's empowerment (see Table 4.11). However, there were significant differences among the four school geographical location groups on all four statements being evaluated.

The P-values of post-hoc tests between all possible combination of school geographical location groups are shown in Table 4.12. On the teachers' involvement in their work, participants from District A reported lower values than participants from the three other districts, and participants from Districts B and C reported higher values than participants from District D. With

Table 4.9 Participants' perceptions of empowerment by number of years in teaching

Item	Years in Teaching	N	M	SD	F	P
Most teachers are highly involved in their work.	<6	380	4.51	0.75	0.56	0.573
	6–15	639	4.48	0.80		
	>15	931	4.52	0.78		
Everyone believes that he or she can have a positive impact.	<6	378	4.29	0.85	0.17	0.842
	6–15	635	4.32	0.84		
	>15	924	4.31	0.91		
Planning is ongoing and involves everyone in the process to some degree	<6	380	4.03	1.03	1.16	0.315
	6–15	634	3.99	1.08		
	>15	923	4.08	1.09		
Information is widely shared so that everyone can get the information he or she needs when it's needed.	<6	380	4.16	1.00	0.90	0.409
	6–15	637	4.22	1.01		
	>15	930	4.24	1.00		
	<6	638	4.37	1.05		
	6–15	931	4.40	0.97		

Table 4.10 Participants' perceptions of empowerment by number of years in teaching: Post-hoc test

Item	<6 vs. 6–15	<6 vs. >15	6–15 vs. >15
Most teachers are highly involved in their work.	0.572	0.774	0.292
Everyone believes that he or she can have a positive impact.	0.588	0.598	0.953
Planning is ongoing and involves everyone in the process to some degree.	0.598	0.475	0.132
Information is widely shared so that everyone can get the information he or she needs when it's needed.	0.305	0.185	0.783

regard to teachers' beliefs on their impact on the school's development, participants from District A reported lower values than participants from the other three districts, and participants from District B reported higher values than participants from Districts C and D. Concerning teachers' opportunities to participate in the school development planning, participants from District A reported lower values than participants from the other three districts. Participants from District B reported higher than participants from Districts

Table 4.11 Participants' perceptions of empowerment by school location

Item	Location	N	M	SD	F	P
Most teachers are highly involved in their work.	District A	357	3.98	1.05	79.10	<0.001
	District B	1011	4.67	0.63		
	District C	132	4.70	0.58		
	District D	486	4.48	0.74		
Everyone believes that he or she can have a positive impact.	District A	354	3.72	0.97	77.21	<0.001
	District B	1002	4.50	0.76		
	District C	133	4.29	0.91		
	District D	484	4.33	0.83		
Planning is ongoing and involves everyone in the process to some degree.	District A	354	3.35	1.13	75.47	<0.001
	District B	1004	4.28	0.98		
	District C	131	3.77	1.20		
	District D	484	4.10	0.99		
Information is widely shared so that everyone can get the information he or she needs when it's needed.	District A	356	3.57	1.09	77.76	<0.001
	District B	1010	4.45	0.88		
	District C	133	4.02	1.07		
	District D	483	4.24	0.95		

C and D. Participants from District C reported lower values than participants from District D. Regarding the availability of and access to school information, participants from District A reported lower values than participants from the other three districts. Participants from District B reported higher values than participants from District C and District D. Participants from District C reported lower values than participants from District D.

4.2 Team Orientation

In order to understand the participants' perceptions of the team orientation of their school culture, four items—"Cooperation across different parts of the school is actively encouraged;" "Sufficient cooperation is taking place among teachers with regard to ethical education, teaching, and other areas;" "Teamwork, rather than hierarchy, is encouraged to get school work done;" "Each teacher can see the relationship between his or her job and the goal of the school"—were asked that required five-point Likert-type responses.

Overall Findings

Table 4.13 shows a summary of the descriptive data of the participants' opinions on these four items. In general, participants reported higher values than the mean on all four items, reporting especially high values on the

Table 4.12 Participants' perceptions of empowerment by location: Post-hoc test

Item	District A vs. B	District A vs. C	District A vs. D	District B vs. C	District B vs. D	District C vs. D
Most teachers are highly involved in their work.	<0.001	<0.001	<0.001	0.653	<0.001	0.002
Everyone believes that he or she can have a positive impact.	<0.001	<0.001	<0.001	0.006	<0.001	0.682
Planning is ongoing and involves everyone in the process to some degree.	<0.001	<0.001	<0.001	<0.001	0.001	0.001
Information is widely shared so that everyone can get the information he or she needs when it's needed.	<0.001	<0.001	<0.001	<0.001	<0.001	0.017

Table 4.13 Participants' perceptions of the school's team orientation

Item	N	M	SD
Cooperation across different parts of the school is actively encouraged.	1989	4.46	0.85
Sufficient cooperation among teachers is taking place in ethical education, teaching, and other areas.	1978	4.39	0.86
Teamwork, is encouraged to get school work done, rather than hierarchy.	1982	4.25	0.98
Each teacher can see the relationship between his or her job and the goal of the school.	1984	4.38	0.84

first one. Participants largely held a positive view of their school's team orientation.

Influence of Gender

Both male and female participants held positive views of their school's team orientation (see Table 4.14). Males and females had similar opinions on cooperating across different parts of the school and encouraging teamwork. Females had more positive opinions on cooperating in ethical education,

Table 4.14 Participants'perceptions of team orientation by gender

Item	Gender	N	M	SD	t	P
Cooperation across different parts of the school is actively encouraged.	Male	353	4.40	0.92	-1.60	0.110
	Female	1620	4.48	0.83		
Sufficient cooperation among teachers is taking place in ethical education, teaching, and other areas.	Male	353	4.31	0.92	-2.00	0.046
	Female	1610	4.41	0.84		
Teamwork is encouraged to get school work done, rather than hierarchy	Male	353	4.22	0.98	-0.75	0.452
	Female	1613	4.26	0.97		
Each teacher can see the relationship between his or her job and the goal of the school.	Male	353	4.25	0.96	–3.31	0.001
	Female	1615	4.41	0.81		

Table 4.15 Participants' perceptions of team orientation by educational level

Item	Education	N	M	SD	F	P
Cooperation across different parts of the school is actively encouraged.	Postgraduate	74	4.39	0.79	0.50	0.680
	Bachelor's degree	1534	4.46	0.85		
	Diploma	284	4.51	0.80		
	Other	35	4.43	0.81		
Sufficient cooperation encouraged among encouraged. teachers is taking place in ethical education, teaching, and other areas.	Postgraduate	74	4.26	0.88	1.14	0.331
	Bachelor's degree	1526	4.38	0.87		
	Diploma	282	4.45	0.79		
	Other	35	4.37	0.73		
Teamwork is encouraged to get school work done, rather than hierarchy.	Postgraduate	73	4.27	0.90	0.78	0.504
	Bachelor's degree	1530	4.24	0.99		
	Diploma	282	4.32	0.93		
	Other	35	4.40	0.69		
Each teacher can see the relationship between his or her job and the goal of the school.	Postgraduate	74	4.22	0.88	2.16	0.091
	Bachelor's degree	1530	4.38	0.84		
	Diploma	283	4.47	0.78		
	Other	35	4.34	0.76		

teaching, and other areas, and teachers' understanding of the relationship between his or her job and the school's development.

Influence of Educational Level

All four educational level groups thought positively of their school's team orientation (see Table 4.15). There were no significant differences among the four educational level groups on these items.

The P-values of post-hoc tests between all possible combinations of educational level groups are shown in Table 4.16.

Influence of Job Duty

All three job duty groups held positive views of their school's team orientation (see Table 4.17). Participants from different job duty groups differed in their perceptions of cooperating across different parts of the school, encouraging teamwork and teachers' understanding of the relationship between their jobs and the school's development. However, there were no significant differences in how the three duty groups viewed the level of cooperation in ethical education, teaching, and other areas.

The P-values of post-hoc tests between job duty groups are shown in Table 4.18. With regard to cooperation across different parts of the school,

Table 4.16 Participants' perceptions of team orientation by educational level: Post-hoc test

Item	Postgraduate vs. Bachelor	Postgraduate vs. Diploma	Postgraduate vs. Other	Bachelor vs. Diploma	Bachelor vs. Other	Diploma vs. Other
Cooperation across different parts of the school is actively encouraged.	0.500	0.281	0.832	0.350	0.830	0.587
Sufficient cooperation among teachers is taking place in ethical education, teaching, and other areas.	0.240	0.084	0.515	0.186	0.971	0.608
Teamwork is encouraged to get school work done, rather than hierarchy.	0.778	0.723	0.528	0.215	0.338	0.642
Each teacher can see the relationship between his or her job and the goal of the school.	0.108	0.018	0.460	0.071	0.817	0.382

Table 4.17 Participants' perceptions of team orientation by job duty

Item	Job duty	N	M	SD	F	P
Cooperation across different parts of the school is actively encouraged.	Teacher	1717	4.46	0.85	3.42	0.033
	Middle leadership	125	4.66	0.71		
	Principal	38	4.42	0.68		
Sufficient cooperation among teachers is taking place in ethical education, teaching, and other areas.	Teacher	1711	4.39	0.86	1.52	0.219
	Middle leadership	123	4.52	0.75		
	Principal	38	4.50	0.56		
Teamwork is encouraged to get school work done, rather than hierarchy.	Teacher	1712	4.26	0.97	4.43	0.012
	Middle leadership	125	4.52	0.87		
	Principal	38	4.34	0.78		
Each teacher can see the relationship between his or her job and the goal of the school.	Teacher	1714	4.38	0.84	5.13	0.006
	Middle leadership	125	4.62	0.66		
	Principal	38	4.32	0.70		

Table 4.18 Participants' perceptions on team orientation by job duty: Post-hoc test

Item	Teacher vs. Middle leadership	Teacher vs. Principal	Middle leadership vs. Principal
Cooperation across different parts of the school is actively encouraged.	0.010	0.757	0.118
Sufficient cooperation among teachers is taking place in ethical education, teaching, and other areas.	0.111	0.446	0.897
Teamwork is encouraged to get school work done, rather than hierarchy.	0.003	0.590	0.319
Each teacher can see the relationship between his or her job and the goal of the school.	0.002	0.626	0.045

teachers reported lower than middle leadership. Concerning the encouragement of teamwork, teacher reported lower than middle leadership. On teachers' understanding of the relationship between his or her job and the school's development, both teachers and principals reported lower than middle leadership.

Influence of Job Title

All three job title groups took a positive view of their school's team orientation (see Table 4.19). Nonetheless, the groups showed statistically significant differences in their responses to all four items. The P-values of post-hoc tests between all possible combinations of job title groups are shown in Table 4.20. Concerning cooperating across different parts of the school, both Junior and Senior groups reported higher values than the Middle group. Regarding cooperating in ethical education, teaching, and other areas, the Senior group reported higher values than the Junior group, which in turn reported higher values than the Middle group. In the area of encouraging teamwork, the Senior group reported higher values than both the Junior and Middle groups. With regard to teachers' understanding of the relationship between their jobs and the school's development, the Senior group reported higher values than both the Junior and Middle groups.

Influence of Number of Years in Teaching

All three groups, those with less than 6 years of teaching experience, 6–15 years of experience, and more than 15 years of teaching experience, had positive views of their school's team orientation (see Table 4.21). There were no significant differences among the three groups with different amounts of experience on cooperating across different parts of the school, cooperating in ethical education, teaching, and other areas, or encouraging teamwork. Groups

Table 4.19 Participants' perceptions on team orientation by job title

Item	Job title	N	M	SD	F	P
Cooperation across different parts of the school is actively encouraged.	Junior	423	4.47	0.83	13.55	<0.001
	Middle	469	4.30	0.90		
	Senior	1000	4.55	0.82		
Sufficient cooperation among teachers is taking place in ethical education, teaching, and other areas.	Junior	423	4.37	0.88	17.98	<0.001
	Middle	464	4.22	0.90		
	Senior	995	4.50	0.81		
Teamwork is encouraged to get school work done, rather than hierarchy.	Junior	422	4.19	1.00	14.69	<0.001
	Middle	466	4.10	0.97		
	Senior	997	4.37	0.95		
Each teacher can see the relationship between his or her job and the goal of the school.	Junior	422	4.32	0.85	19.71	<0.001
	Middle	465	4.23	0.85		
	Senior	1000	4.51	0.80		

Table 4.20 Participants' perceptions of team orientation by job title: Post-hoc test

Item	Junior vs. Middle	Junior vs. Senior	Middle vs. Senior
Cooperation across different parts of the school is actively encouraged.	0.004	0.084	<0.001
Sufficient cooperation among teachers is taking place in ethical education, teaching, and other areas.	0.009	0.007	<0.001
Teamwork is encouraged to get school work done, rather than hierarchy.	0.163	0.001	<0.001
Each teacher can see the relationship between his or her job and the goal of the school.	0.115	<0.001	<0.001

Table 4.21 Participants' perceptions of team orientation by number of years in teaching

Item	Years in Teaching	N	M	SD	F	P
Cooperation across different parts of the school is actively encouraged.	<6	382	4.44	0.81	2.75	0.064
	6–15	639	4.41	0.87		
	>15	932	4.51	0.84		
Sufficient cooperation among teachers is taking place in ethical education, teaching, and other areas.	<6	380	4.33	0.88	1.47	0.229
	6–15	637	4.38	0.86		
	>15	926	4.42	0.85		
Teamwork is encouraged to get school work done, rather than hierarchy.	<6	380	4.16	0.98	2.94	0.053
	6–15	638	4.27	0.94		
	>15	928	4.30	0.99		
Each teacher can see the relationship between his or her job and the goal of the school.	<6	381	4.27	0.86	5.10	0.006
	6–15	636	4.40	0.82		
	>15	931	4.43	0.83		

with different amounts of experience differed in their perceptions of the understanding of the relationship between their jobs and the school's development.

The P-values of post-hoc tests between all possible combinations of years in teaching groups are shown in Table 4.22. Concerning teachers'

Table 4.22 Participants' perceptions of team orientation by years in teaching: Post-hoc test

Item	<6 vs. 6–15	<6 vs. >15	6–15 vs. >15
Cooperation across different parts of the school is actively encouraged.	0.625	0.166	0.024
Sufficient cooperation among teachers is taking place in ethical education, teaching, and other areas.	0.354	0.090	0.401
Teamwork is encouraged to get school work done, rather than hierarchy.	0.084	0.016	0.493
Each teacher can see the relationship between his or her job and the goal of the school.	0.013	0.002	0.546

understanding of the relationship between their jobs and the school's development, the group with less than six years of experience reported lower values than the group with 6–15 years of experience and the group with more than 15 years.

Influence of the School's Geographical Location

All four school geographical location groups felt positively about their school's team orientation (see Table 4.23). However, there were significant differences among the four school geographical location groups on all four items being evaluated.

The P-values of post-hoc tests between all possible combinations of school geographical location groups are shown in Table 4.24. Concerning cooperating across different parts of the school, participants from District A reported lower values than participants from the other three districts. Participant from District B reported higher values than participants from Districts C and D. Concerning cooperating in ethical education, teaching, and other areas, participants from District A reported lower values than participants from the other three districts. Participants from District B reported higher values than participants from Districts C and D. Concerning encouraging teamwork, participants from District A reported lower values than participants from the other districts. Participants from District B reported higher values than participants from Districts C and D. Participants from District C reported lower values than participants from District D. With regard to teachers' understanding of the relationship between their jobs and the school's development, participants from District A reported lower values than participants from the other districts. Participants from District B reported higher values than participants from Districts C and D.

Table 4.23 Participants' perceptions of team orientation by location

Item	Location	N	M	SD	F	P
Cooperation across different parts of the school is actively encouraged.	District A	358	3.76	1.06	124.21	<0.001
	District B	1011	4.69	0.63		
	District C	133	4.45	0.85		
	District D	487	4.51	0.79		
Sufficient cooperation among teachers is taking place in ethical education, teaching, and other areas.	District A	354	3.71	1.01	111.48	<0.001
	District B	1008	4.61	0.67		
	District C	133	4.39	0.91		
	District D	483	4.41	0.83		
Teamwork is encouraged to get school work done, rather than hierarchy.	District A	356	3.65	1.07	75.37	<0.001
	District B	1008	4.49	0.84		
	District C	133	3.98	1.09		
	District D	485	4.28	0.95		
Each teacher can see the relationship between his or her job and the goal of the school.	District A	355	3.76	1.02	107.86	<0.001
	District B	1011	4.62	0.63		
	District C	133	4.32	0.88		
	District D	485	4.35	0.82		

Table 4.24 Participants' perceptions of team orientation by location: Post-hoc test

Item	District A vs. B	District A vs. C	District A vs. D	District B vs. C	District B vs. D	District C vs. D
Cooperation across different parts of the school is actively encouraged.	<0.001	<0.001	<0.001	0.001	<0.001	0.479
Sufficient cooperation among teachers is taking place in ethical education, teaching, and other areas.	<0.001	<0.001	<0.001	0.003	<0.001	0.808
Teamwork is encouraged to get school work done, rather than hierarchy.	<0.001	<0.001	<0.001	<0.001	<0.001	0.001
Each teacher can see the relationship between his or her job and the goal of the school.	<0.001	<0.001	<0.001	<0.001	<0.001	0.630

4.3 Capability Development

In order to understand participants' perceptions of the capability development of their school culture, four items—"Authority is delegated so that teachers can act on their own;" "The professional skills of teachers are constantly improving;" "There is continuous financial investment from the school to improve the professional skills of the teachers;" "Problems often arise because some teachers do not have the skills necessary to do the job"—were asked that required five-point Likert-type responses.

Overall Findings

Table 4.25 shows a summary of the descriptive data of participants' responses to these items. In general, participants reported higher values than the mean on all four of these items, especially on the second item. Participants generally held a positive opinion of their school's capability development.

Influence of Gender

Both male and female participants held positive opinions of their school's capability development (see Table 4.26). Males and females had similar opinions on the delegation of teachers' authority and the continuous financial investment made from the school to improve teachers' professional skills. Females agreed more strongly with statements regarding improving teachers' professional skill and the problems caused by teachers lacking necessary skills.

Influence of Educational Level

All four educational level groups held positive opinions of their school's capability development (see Table 4.27). There were no significant differences among the educational level groups on delegating authority to teachers, improving teachers' professional skills, and the continuous

Table 4.25 Participants' perceptions of their school's capability development

Item	N	M	SD
Authority is delegated so that teachers can act on their own.	1978	3.70	1.18
The professional skills of teachers are constantly improving.	1975	4.44	0.81
There is continuous financial investment from the school to improve the professional skills of the teachers.	1971	4.24	1.00
Problems often arise because some teachers do not have the skills necessary to do the job.	1986	4.34	1.05

Table 4.26 Participants' perceptions of capability development by gender

Item	Gender	N	M	SD	t	P
Authority is delegated so that teachers can act on their own.	Male	353	3.70	1.14	−0.12	0.906
	Female	1609	3.71	1.19		
The professional skills of teachers are constantly improving.	Male	351	4.36	0.89	−2.05	0.041
	Female	1610	4.46	0.79		
There is continuous financial investment from the school to improve the professional skills of the teachers.	Male	350	4.23	1.08	−0.36	0.721
	Female	1606	4.25	0.99		
Problems often arise because some teachers do not have the skills necessary to do the job.	Male	353	4.00	1.27	−6.73	<0.001
	Female	1617	4.41	0.98		

Table 4.27 Participants' perceptions of capability development by educational level

Item	Education	N	M	SD	F	P
Authority is delegated so that teachers can act on their own.	Postgraduate	74	4.00	1.02	1.71	0.163
	Bachelor's degree	1526	3.69	1.20		
	Diploma	280	3.71	1.13		
	Other	36	3.78	1.02		
The professional skills of teachers are constantly improving.	Postgraduate	74	4.35	0.73	2.04	0.106
	Bachelor's degree	1524	4.43	0.83		
	Diploma	283	4.54	0.72		
	Other	35	4.49	0.82		
There is continuous financial investment from the school to improve the professional skills of the teachers.	Postgraduate	73	4.25	0.85	1.30	0.274
	Bachelor's degree	1523	4.23	1.01		
	Diploma	280	4.34	0.95		
	Other	34	4.38	1.02		
Problems often arise because some teachers do not have the skills necessary to do the job.	Postgraduate	74	3.41	1.46	21.87	<0.001
	Bachelor's degree	1531	4.36	1.01		
	Diploma	283	4.46	0.99		
	Other	36	4.19	1.17		

financial investment from the school to improve teachers' professional skills. Participants from different educational level groups differed in their perceptions of the problems caused by teachers lacking necessary skills.

The P-values of post-hoc tests between all possible combinations of educational level groups are shown in Table 4.28. With regard to problems being caused due to teachers' lack of necessary skills, participants with postgraduate education reported lower values than the other three educational level groups.

Influence of Job Duty

All three job duty groups held positive views of their school's capability development (see Table 4.29). Participants from different job duty groups reported similar in their perceptions of delegating authority to teachers, improving teachers' professional skills, and having problems due to teachers lacking necessary skills. Nonetheless, there were significant differences in opinion among job duty groups regarding the continuous financial investments from the school to improve teachers' professional skills.

Table 4.28 Participants' perceptions of capability development by educational level: Post-hoc test

Item	Postgraduate vs. Bachelor	Postgraduate vs. Diploma	Postgraduate vs. Other	Bachelor vs. Diploma	Bachelor vs. Other	Diploma vs. Other
Authority is delegated so that teachers can act on their own.	0.026	0.064	0.354	0.726	0.650	0.761
The professional skills of teachers are constantly improving.	0.434	0.068	0.417	0.024	0.668	0.686
There is continuous financial investment from the school to improve the professional skills of the teachers.	0.863	0.463	0.512	0.072	0.366	0.828
Problems often arise because some teachers do not have the skills necessary to do the job.	<0.001	<0.001	<0.001	0.137	0.342	0.147

Table 4.29 Participants' perceptions of capability development by job duty

Item	Job duty	N	M	SD	F	P
Authority is delegated so that teachers can act on their own.	Teacher	1711	3.72	1.19	0.35	0.703
	Middle leadership	122	3.76	1.05		
	Principal	38	3.58	0.95		
The professional skills of teachers are constantly improving.	Teacher	1709	4.44	0.81	2.24	0.107
	Middle leadership	124	4.59	0.69		
	Principal	38	4.55	0.50		
There is continuous financial investment from the school to improve the professional skills of the teachers.	Teacher	1703	4.22	1.02	14.33	<0.001
	Middle leadership	124	4.65	0.68		
	Principal	38	4.68	0.66		
Problems often arise because some teachers do not have the skills necessary to do the job.	Teacher	1716	4.35	1.05	0.08	0.924
	Middle leadership	125	4.34	0.99		
	Principal	38	4.29	0.84		

The P-values of post-hoc tests between all possible combinations of job duty groups are shown in Table 4.30. Regarding the continuous financial investments from the school to improve teachers' professional skills, teachers reported lower values than middle leadership and principals.

Influence of Job Title

All three job title groups held positive perceptions of their school's capability development (see Table 4.31). There were no significant differences among the three groups on delegating authority to teachers. However, groups with different job titles differed in their perceptions of the improvement of teachers' professional skills, the continuous financial investment from the school to improve teachers' professional skills, and the problems which are due to teachers not having the necessary skills.

The P-values of post-hoc tests between all possible combinations of job title groups are shown in Table 4.32. Regarding the improvement of teachers' professional skills, the Senior group reported higher values than the Junior group, which in turn reported higher values than the Middle group. Concerning the continuous financial investment from the school to improve teachers' professional skills, both the Junior and Middle groups reported lower values than the Senior group. Regarding problems caused by teachers' lack of necessary skills, both the Junior and Middle groups reported lower values than the Senior group.

Table 4.30 Participants' perceptions of capability development by job duty: Post-hoc test

Item	Teacher vs. Middle leadership	Teacher vs. Principal	Middle leadership vs. Principal
Authority is delegated so that teachers can act on their own.	0.706	0.463	0.402
The professional skills of teachers are constantly improving.	0.049	0.399	0.808
There is continuous financial investment from the school to improve the professional skills of the teachers.	<0.001	0.004	0.832
Problems often arise because some teachers do not have the skills necessary to do the job.	0.868	0.714	0.809

Table 4.31 Participants' perceptions of their school's capability development by job title

Item	Job Title	N	M	SD	F	P
Authority is delegated so that teachers can act on their own.	Junior	421	3.73	1.17	2.80	0.061
	Middle	465	3.61	1.15		
	Senior	995	3.76	1.21		
The professional skills of teachers are constantly improving.	Junior	424	4.40	0.84	19.10	<0.001
	Middle	462	4.27	0.85		
	Senior	995	4.54	0.77		
There is continuous financial investment from the school to improve the professional skills of the teachers.	Junior	420	4.16	1.05	24.26	<0.001
	Middle	460	4.03	1.03		
	Senior	996	4.40	0.94		
Problems often arise because some teachers do not have the skills necessary to do the job.	Junior	424	4.28	1.08	15.94	<0.001
	Middle	466	4.15	1.18		
	Senior	999	4.47	0.95		

Influence of Number of Years in Teaching

All three groups, those with less than 6 years of teaching experience, 6–15 years of experience, and more than 15 years of teaching experience, had positive views of their school's team orientation capability

Table 4.32 Participants' perceptions of the school's capability development by job title: Post-hoc test

Item	Junior vs. Middle	Junior vs. Senior	Middle vs. Senior
Authority is delegated so that teachers can act on their own.	0.117	0.650	0.019
The professional skills of teachers are constantly improving.	0.014	0.002	<0.001
There is continuous financial investment from the school to improve the professional skills of the teachers.	0.057	<0.001	<0.001
Problems often arise because some teachers do not have the skills necessary to do the job.	0.062	0.002	<0.001

development (see Table 4.33). There were no significant differences among three groups with different levels of experience on their perceptions of delegating authority to teachers and improving teachers' professional skills. Groups with different levels of experience differed in their perceptions of the continuous financial investment from the school to improve teachers' professional skills and the problems caused by teachers lacking necessary skills.

The P-values of post-hoc tests between all possible combinations of years in teaching groups are shown in Table 4.34. With regard to the continuous financial investment from the school to improve teachers' professional skills, the group with less than six year of experience and the group with 6–15 years of experience reported lower values than the group with more than 15 years of experience. Concerning the problems caused by teachers lacking necessary skills, the group with less than 6 years of experience reported lower values than the group with 6–15 or more than 15 years of experience.

Influence of the School's Geographical Location

All four school geographical location groups held positive views of their school's capability development (see Table 4.35). However, there were significant differences among the four groups on all four items being evaluated.

The P-values of post-hoc tests between all possible combinations of school geographical location groups are shown in Table 4.36. With regard to teachers' ability to act on their own and the improvement of teachers' professional

Table 4.33 Participants' perceptions of capability development by years in teaching

Item	Years in Teaching	N	M	SD	F	P
Authority is delegated so that teachers can act on their own.	<6	378	3.71	1.16	0.00	0.997
	6–15	638	3.71	1.17		
	>15	926	3.71	1.20		
The professional skills of teachers are constantly improving.	<6	380	4.37	0.82	2.64	0.072
	6–15	634	4.43	0.79		
	>15	927	4.48	0.82		
There is continuous financial investment from the school to improve the professional skills of the teachers.	<6	378	4.12	1.01	6.85	0.001
	6–15	634	4.20	1.01		
	>15	924	4.33	0.98		
Problems often arise because some teachers do not have the skills necessary to do the job.	<6	381	4.14	1.19	8.84	<0.001
	6–15	638	4.37	1.05		
	>15	931	4.40	0.97		

Table 4.34 Participants' perceptions of the school's capability development by years in teaching: Post-hoc test

Item	<6 vs. 6–15	<6 vs. >15	6–15 vs. >15
Authority is delegated so that teachers can act on their own.	0.945	0.953	0.987
The professional skills of teachers are constantly improving.	0.212	0.023	0.266
There is continuous financial investment from the school to improve the professional skills of the teachers.	0.262	0.001	0.010
Problems often arise because some teachers do not have the skills necessary to do the job.	0.001	<0.001	0.548

skills, participants from District A reported lower values than participants from the other three districts. For the improvement of teachers' professional skills, participants from District A reported lower values than participants from the other three districts, and participants from District B reported higher values than participants from District D. Concerning the continuous

Table 4.35 Participants' perceptions of the school's capability development by location

Item	Location	N	M	SD	F	P
Authority is delegated so that teachers can act on their own.	District A	356	3.23	1.20	25.96	<0.001
	District B	1006	3.85	1.16		
	District C	133	3.64	1.26		
	District D	483	3.78	1.09		
The professional skills of teachers are constantly improving.	District A	352	3.80	1.02	110.62	<0.001
	District B	1007	4.64	0.63		
	District C	133	4.55	0.74		
	District D	483	4.45	0.76		
There is continuous financial investment from the school to improve the professional skills of the teachers.	District A	351	3.50	1.14	95.72	<0.001
	District B	1007	4.48	0.86		
	District C	131	4.18	1.05		
	District D	482	4.31	0.90		
Problems often arise because some teachers do not have the skills necessary to do the job.	District A	356	4.05	1.02	37.63	<0.001
	District B	1011	4.57	0.86		
	District C	133	4.29	1.00		
	District D	486	4.07	1.29		

Table 4.36 Participants' perceptions of the school's capability development by location: Post-hoc test

Item	District A vs. B	District A vs. C	District A vs. D	District B vs. C	District B vs. D	District C vs. D
Authority is delegated so that teachers can act on their own.	<0.001	<0.001	<0.001	0.052	0.286	0.220
The professional skills of teachers are constantly improving.	<0.001	<0.001	<0.001	0.173	<0.001	0.177
There is continuous financial investment from the school to improve the professional skills of the teachers.	<0.001	<0.001	<0.001	0.001	0.001	0.180
Problems often arise because some teachers do not have the skills necessary to do the job.	<0.001	0.021	0.771	0.003	<0.001	0.028

financial investments of the school to improve teachers' professional skills, participants from District A reported lower values than participants from the other three districts. Participants from District B reported higher values than participants from Districts C and D. Concerning the problems caused by teachers lacking necessary skills, participants from District A reported lower values than participants from Districts B and C, and participants from District B reported higher values than participants from Districts C and D, participants from District C reported higher values than participants from District D.

4.4 Summary

The purpose of this chapter was to investigate school staff members' perceptions of involvement, one of the school culture traits. There have been three sections in this chapter, one for each of the three indexes of involvement: empowerment, team orientation, and capability development. Four items were used for each area, using a five-point Likert-type item, which the participants then responded to. First, participants' ideas on empowerment are presented. Four items are used: Most teachers are highly involved in their work; Everyone believes that he or she can have a positive impact; Planning is ongoing and involves everyone in the process to some degree; Information is widely shared so that everyone can get the information he or she needs when it's needed. Second, participants' ideas on team orientation are presented. Four items are used: Cooperation across different parts of the school is actively encouraged; Sufficient cooperation is taking place among teachers with regard to ethical education, teaching, and other areas; Teamwork is encouraged to get school work done, rather than hierarchy; Each teacher can see the relationship between his or her job and the goal of the school. Third, participants report their opinions on capability development. There are four items in this section: Authority is delegated so that teachers can act on their own; The professional skills of teachers are constantly improving; There is continuous financial investment from the school to improve the professional skills of the teachers; Problems often arise because some teachers do not have the skills necessary to do the job.

The overall findings of the participants' responses to the items were recorded. The effects of variables such as gender, educational level, job duty, job title, years in teaching, and school location were considered for each item. The findings presented in this chapter can be summarized as follows:

1. Participants in this investigation generally agreed on the empowerment of their school culture. They believed most teachers in their school are highly involved in their work and that everyone in their schools believes that he or she can make positive impacts. They agreed that planning is ongoing and involves everyone, to some degree, in the process. They felt that information is widely shared and they can obtain information when needed. However, significant differences can be observed between some groups in their perceptions of school's empowerment; for example, the Middle group reported a lower degree of agreement than the Senior group, and participants from District A reported a lower degree of agreement than participants from other three districts. The factors of years in teaching and educational level did not have any effect on this index.
2. Findings from the team orientation section showed a positive overall picture of staff members' agreement on this school culture index. Participants agreed that management actively encouraged cooperation across different parts of the school. They thought sufficient cooperation was taking place among teachers in ethical education, teaching, and other areas. They also thought that teamwork was encouraged. They thought each teacher understood the relationship between his or her job and the school's goals. The factor of educational level did not have an effect on team orientation. However, differences can be found among job title groups and school location groups. Concerning the effect of job title, the Middle group reported a lower degree of agreement than the Senior group on their perception of school's team orientation. Participants from District A reported a lower degree of agreement than participants from other three districts, and participants from District B reported higher levels of agreement than participants from Districts C and D on their perception of school's team orientation.
3. Generally, participants agreed on the capability development of their school's culture. They thought authority was delegated to the teachers and that teachers' professional skills were constantly being improved. They agreed that their schools provided continuous financial investment to improve teachers' professional skills. They also agreed that there were few problems due to teachers' lack of necessary skills. Nevertheless, there was a significant difference among school location groups on their perception of school's capability development; that is participants from District A reported a lower degree of agreement than participants from Districts B and C.

In conclusion, the participants strongly agreed that their schools have good cultures concerning involvement. The extent of agreement differed among groups based on gender and job duty. Furthermore, school location and the staff members' job titles played an important role in the perceptions of their school's involvement. The results also indicate further need to enhance involvement for participants from District A and participants in the Middle job title group.

5

Consistency

This chapter deals with school staff members' perceptions of one of the school culture traits, *consistency*. Consistency consists of three indexes, namely *core values*, *team orientation*, and *coordination and integration*. Twelve survey items were used as a way to evaluate these indexes. The findings are presented, analyzed in three sections. Section one reports how participants responded to questions about their awareness of the principal's educational philosophy, understanding of the school's development goals, understanding of the school motto's cultural meaning, and the consistency of their mode of dress and behavior with school culture. Section two presents participants' perceptions of the hard working to achieve "win-win" solutions when disagreements occur, their rates of success in reaching consensus, the existence of a clear agreement about the right and wrong way to do things, and the trouble of reaching agreement on key issues. Section three shows how participants responded to the existence of a common perspective among people from different parts of the school, the success of coordination across different parts of the school, the challenges of cooperation in the school, and the alignment of goals across all levels. Each section details the overall findings for the given items as well as the effects of participants' gender, educational level, job duty, job title, years in teaching, and school location on the findings.

5.1 Core Values

In order to understand participants' perceptions of the core values of their school culture, four questions–"I am aware of the principal's educational philosophy and can adjust my own accordingly;" "I understand the school's development goals;" "I understand the school motto's cultural meaning;" "My mode of dress and behavior are consistent with school culture"—were used that required five-point Likert-type responses.

Overall Findings

A summary of the descriptive data of participants' responses to these items is shown in Table 5.1. In general, participants reported higher values than the mean on all items, especially on the last two. Participants held a positive perception of their school's core values.

Influence of Gender

Both male and female participants took a favorable view of their school's core values (see Table 5.2). Males and females gave similar assessments of their awareness of the principal's educational philosophy and understanding of the school's development goals. In general, females were more likely to understand the school motto's cultural meaning and to feel that their mode of dress and behavior were consistent with school culture.

Influence of Educational Level

All four educational level groups had positive perceptions of their school's core values (see Table 5.3). There was no significant difference among the four educational level groups on any of the items.

The P-values of post-hoc tests between all possible combinations of two educational level groups are shown in Table 5.4.

Influence of Job Duty

All three job title groups held positive attitudes towards their school's core values (see Table 5.5). Teachers, middle leadership administrators, and principals had different levels of awareness of all four items.

The P-values of post-hoc tests between all possible combinations of educational level groups are shown in Table 5.6. Teachers reported lower values than middle leadership on the their awareness of the principal's educational philosophy and understanding of the school's development goals. Concerning the understanding of the school motto's cultural meaning, both teachers and principals reported lower values than middle leadership. Regarding the

Table 5.1 Participants' perceptions of core values

Item	N	M	SD
I am aware of the principal's educational philosophy and can adjust my own accordingly.	1983	4.37	0.83
I understand the school's development goals.	1986	4.40	0.82
I understand the school motto's cultural meaning.	1983	4.45	0.79
My mode of dress and behavior are consistent with school culture.	1946	4.45	0.76

Table 5.2 Participants' perceptions of core values by gender

Item	Gender	N	M	SD	t	P
I am aware of the principal's educational philosophy and can adjust my own accordingly.	Male	353	4.34	0.81	–0.75	0.452
	Female	1615	4.38	0.83		
I understand the school's development goals.	Male	354	4.34	0.83	−1.53	0.127
	Female	1616	4.41	0.82		
I understand the school motto's cultural meaning.	Male	353	4.38	0.81	−1.97	0.049
	Female	1614	4.47	0.78		
My mode of dress and behavior are consistent with school culture.	Male	349	4.36	0.81	–2.57	0.010
	Female	1582	4.47	0.75		

Table 5.3 Participants' perceptions of core values by educational level

Item	Education	N	M	SD	F	P
I am aware of the principal's educational philosophy and can adjust my own accordingly.	Postgraduate	74	4.36	0.84	0.06	0.980
	Bachelor's degree	1531	4.37	0.82		
	Diploma	282	4.35	0.83		
	Other	35	4.34	0.76		
I understand the school's development goals.	Post-graduate	74	4.31	0.81	0.86	0.461
	Bachelor's degree	1531	4.40	0.82		
	Diploma	283	4.43	0.81		
	Other	36	4.25	0.97		
I understand the school motto's cultural meaning.	Postgraduate	74	4.35	0.78	1.31	0.271
	Bachelor's degree	1529	4.44	0.79		
	Diploma	283	4.52	0.75		
	Other	35	4.54	0.66		
My mode of dress and behavior are consistent with school culture.	Postgraduate	73	4.27	0.79	1.57	0.195
	Bachelor's degree	1501	4.46	0.76		
	Diploma	275	4.46	0.76		
	Other	35	4.34	0.87		

consistency of their mode of dress and behavior with school culture, both teachers and middle leadership reported higher values than principals.

Influence of Job Title

All three job title groups felt positively about their school's core values (see Table 5.7), though there were significant differences among the three job title groups on all four items.

Table 5.4 Participants' perceptions of core values by educational level: Post-hoc test

Item	Postgraduate vs. Bachelor	Postgraduate vs. Diploma	Postgraduate vs. Other	Bachelor vs. Diploma	Bachelor vs. Other	Diploma vs. Other
I am aware of the principal's educational philosophy and can adjust my own accordingly.	0.945	0.898	0.896	0.700	0.838	0.956
I understand the school's development goals.	0.376	0.247	0.715	0.479	0.287	0.203
I understand the school motto's cultural meaning.	0.322	0.102	0.235	0.139	0.462	0.868
My mode of dress and behavior are consistent with school culture.	0.046	0.067	0.661	0.982	0.383	0.401

Table 5.5 Participants' perceptions of core values by job duty

Item	Job duty	N	M	SD	F	P
I am aware of the principal's educational philosophy and can adjust my own accordingly.	Teacher	1715	4.37	0.83	4.95	0.007
	Middle leadership	125	4.60	0.62		
	Principal	38	4.32	0.66		
I understand the school's development goals.	Teacher	1716	4.40	0.82	5.00	0.007
	Middle leadership	125	4.63	0.65		
	Principal	38	4.37	0.63		
I understand the school motto's cultural meaning.	Teacher	1713	4.46	0.79	3.64	0.027
	Middle leadership	125	4.63	0.68		
	Principal	38	4.32	0.74		
My mode of dress and behavior are consistent with school culture.	Teacher	1679	4.47	0.76	4.78	0.009
	Middle leadership	124	4.53	0.69		
	Principal	38	4.11	0.76		

The P-values of post-hoc tests between all possible combinations of two job title groups are shown in Table 5.8. For all four items, the Senior group reported higher values than the Junior group, which in turn reported higher values than the Middle group.

Table 5.6 Participants' perceptions of core values by job duty: Post-hoc test

Item	Teacher vs. Middle leadership	Teacher vs. Principal	Middle leadership vs. Principal
I am aware of the principal's educational philosophy and can adjust my own accordingly.	0.002	0.713	0.060
I understand the school's development goals.	0.002	0.826	0.077
I understand the school motto's cultural meaning.	0.016	0.267	0.029
My mode of dress and behavior are consistent with school culture.	0.345	0.004	0.002

Table 5.7 Participants' perceptions of core values by job title

Item	Job title	N	M	SD	F	P
I am aware of the principal's educational philosophy and can adjust my own accordingly.	Junior	423	4.33	0.84	25.21	<0.001
	Middle	466	4.18	0.87		
	Senior	998	4.49	0.77		
I understand the school's development goals.	Junior	423	4.34	0.85	37.20	<0.001
	Middle	466	4.18	0.89		
	Senior	1000	4.55	0.71		
I understand the school motto's cultural meaning.	Junior	423	4.39	0.80	30.03	<0.001
	Middle	466	4.27	0.87		
	Senior	997	4.59	0.69		
My mode of dress and behavior are consistent with school culture.	Junior	416	4.45	0.74	21.55	<0.001
	Middle	454	4.27	0.82		
	Senior	979	4.55	0.72		

Table 5.8 Participants' perceptions of core values by job title: Post-hoc test

Item	Junior vs. Middle	Junior vs. Senior	Middle vs. Senior
I am aware of the principal's educational philosophy and can adjust my own accordingly.	0.007	<0.001	<0.001
I understand the school's development goals.	0.004	<0.001	<0.001
I understand the school motto's cultural meaning.	0.022	<0.001	<0.001
My mode of dress and behavior are consistent with school culture.	<0.001	0.022	<0.001

Influence of Number of Years in Teaching

All three of the teaching groups, those with less than 6 years of teaching experience, 6–15 years of experience, and more than 15 years of teaching experience, held positive opinions of their school's core values (see

Table 5.9). There were no significant differences among the three teaching groups on their awareness of the principal's educational philosophy or the consistency of their mode of dress and behavior with school culture. However, participants with different amounts of teaching experience had differing understandings of the school's development goals and of the school motto's cultural meaning.

The P-values of post-hoc tests between the various groups according to the number of years in teaching are shown in Table 5.10. Concerning the understanding of the school's development goals, the group with less than 6 years of teaching experiences reported lower than the groups with 6–15 years of teaching experiences and more than 15 years of teaching experience. With regard to the understanding of the school motto, the groups with less than 6 years and 6–15 years of teaching experiences reported lower than the group with more than 15 years of teaching experiences.

Influence of the School's Geographical Location

All four school geographical location groups had positive views of their school's core values (see Table 5.11). However, there were significant differences among the four school geographical location groups on all four statements being evaluated.

The P-values of post-hoc tests between all possible combinations of school geographical location groups are shown in Table 5.12. On their awareness of

Table 5.9 Participants' perceptions of core values by years in teaching

Item	Years in Teaching	N	M	SD	F	P
I am aware of the principal's educational philosophy and can adjust my own accordingly.	<6	381	4.33	0.83	1.81	0.164
	6–15	638	4.35	0.80		
	>15	929	4.41	0.83		
I understand the school's development goals.	<6	381	4.28	0.86	6.06	0.002
	6–15	638	4.39	0.82		
	>15	931	4.45	0.79		
I understand the school motto's cultural meaning.	<6	381	4.36	0.80	5.49	0.004
	6–15	638	4.43	0.80		
	>15	928	4.51	0.76		
My mode of dress and behavior are consistent with school culture.	<6	375	4.41	0.76	1.60	0.202
	6–15	627	4.44	0.75		
	>15	909	4.49	0.76		

Table 5.10 Participants' perceptions of core values by years in teaching: Post-hoc test

Item	<6 vs. 6–15	<6 vs. >15	6–15 vs. >15
I am aware of the principal's educational philosophy and can adjust my own accordingly.	0.730	0.104	0.137
I understand the school's development goals.	0.049	0.001	0.108
I understand the school motto's cultural meaning.	0.165	0.002	0.046
My mode of dress and behavior are consistent with school culture.	0.517	0.094	0.244

Table 5.11 Participants' perceptions of core values by school location

Item	Location	N	M	SD	F	P
I am aware of the principal's educational philosophy and can adjust my own accordingly.	District A	356	3.83	0.95	83.27	<0.001
	District B	1010	4.58	0.67		
	District C	133	4.30	0.86		
	District D	484	4.34	0.82		
I understand the school's development goals.	District A	357	3.75	0.96	125.52	<0.001
	District B	1010	4.64	0.63		
	District C	133	4.32	0.83		
	District D	486	4.38	0.78		
I understand the school motto's cultural meaning.	District A	355	3.91	0.92	92.19	<0.001
	District B	1009	4.66	0.63		
	District C	133	4.40	0.83		
	District D	486	4.43	0.77		
My mode of dress and behavior are consistent with school culture.	District A	353	4.04	0.89	50.21	<0.001
	District B	976	4.60	0.67		
	District C	133	4.45	0.80		
	District D	484	4.43	0.73		

the principal's educational philosophy, participants from District A reported lower values than participants from the other three districts, and participants from District B reported higher values than participants from Districts C and D. With regard to understanding of the school's development goals, participants from District A reported lower values than participants from the other three districts, and participants from District B reported higher values than participants from Districts C and D. Concerning understanding of the school motto's cultural meaning, participants from District A reported lower values than participants from the other three districts. Participants from District B reported higher values than participants from Districts C and D. Regarding the consistency of their mode of dress and behavior with school culture, participants from District A reported lower values than participants

Table 5.12 Participants' perceptions of core values by location: Post-hoc test

Item	District A vs. B	District A vs. C	District A vs. D	District B vs. C	District B vs. D	District C vs. D
I am aware of the principal's educational philosophy and can adjust my own accordingly.	<0.001	<0.001	<0.001	<0.001	<0.001	0.637
I understand the school's development goals.	<0.001	<0.001	<0.001	<0.001	<0.001	0.419
I understand the school motto's cultural meaning.	<0.001	<0.001	<0.001	<0.001	<0.001	0.663
My mode of dress and behavior are consistent with school culture.	<0.001	<0.001	<0.001	0.025	<0.001	0.811

from the other three districts. Participants from District B reported higher values than participants from Districts C and D. There were no differences between participants from Districts C and D on these four items.

5.2 Agreement

In order to understand the participants' perceptions of agreement in their school culture, four items—"When disagreements occur, we work hard to achieve 'win-win solutions;" "It is easy to reach consensus, even on difficult issues;" "There is clear agreement about the right and wrong way to do things;" "We often have trouble reaching agreement on key issues"—were asked that required 5-point Likert-type responses.

Overall Findings

Table 5.13 shows a summary of the descriptive data of participants' opinions on these four items. In general, participants reported higher values than the mean on all four items, reporting especially high values on the second one.

Influence of Gender

Both male and female participants held positive views of their school's agreement (see Table 5.14). Males and females had similar opinions on

Table 5.13 Participants' perceptions of agreement

Item	N	M	SD
When disagreements occur, we work hard to achieve "win-win" solutions.	1985	4.23	0.95
It is easy to reach consensus, even on difficult issues.	1981	4.32	0.85
There is a clear agreement about the right way and the wrong way to do things.	1981	4.23	0.93
We often have trouble reaching agreement on key issues.	1980	4.26	1.17

Table 5.14 Participants' perceptions of agreement by gender

Item	Gender	N	M	SD	t	P
When disagreements occur, we work hard to achieve "win-win" solutions.	Male	351	4.15	1.01	−1.76	0.078
	Female	1618	4.25	0.93		
It is easy to reach consensus, even on difficult issues.	Male	353	4.21	0.90	−2.70	0.007
	Female	1612	4.34	0.84		
There is a clear agreement about the right way and the wrong way to do things.	Male	353	4.19	0.97	−1.11	0.265
	Female	1612	4.25	0.92		
We often have trouble reaching agreement on key issues.	Male	351	4.00	1.29	−4.59	<0.001
	Female	1613	4.32	1.13		

the hard working to achieve "win-win" solutions when disagreements occur and the existence of clear agreement about the right and wrong way to do things. Females had more positive opinions than males on the rate of success in reaching consensus and the trouble of reaching agreement on key issues.

Influence of Educational Level

All four educational level groups thought positively of their school's agreement (see Table 5.15). There was no significant difference among the four educational level groups on the hard working to achieve "win-win" solutions when disagreements occur, the rate of success in reaching consensus, the existence of clear agreement about the right and wrong way to do things, or the trouble of reaching agreement on key issues.

The *P*-values of post-hoc tests between all possible combinations of educational level groups are shown in Table 5.16.

Table 5.15 Participants' perceptions of agreement by educational level

Item	Education	N	M	SD	F	P
When disagreements occur, we work hard to achieve "win-win" solutions.	Postgraduate	74	4.30	0.84	0.17	0.913
	Bachelor's degree	1532	4.23	0.95		
	Diploma	283	4.21	0.97		
	Other	35	4.23	0.91		
It is easy to reach consensus, even on difficult issues.	Postgraduate	73	4.23	0.86	0.74	0.528
	Bachelor's degree	1531	4.31	0.85		
	Diploma	281	4.35	0.86		
	Other	34	4.47	0.71		
There is a clear agreement about the right way and the wrong way to do things.	Postgraduate	74	4.23	0.84	0.02	0.995
	Bachelor's degree	1527	4.23	0.93		
	Diploma	283	4.24	0.91		
	Other	35	4.26	0.92		
We often have trouble reaching agreement on key issues.	Postgraduate	74	4.04	1.19	2.13	0.095
	Bachelor's degree	1529	4.25	1.18		
	Diploma	280	4.38	1.10		
	Other	35	4.09	1.29		

Table 5.16 Participants' perceptions of agreement by educational level: Post-hoc test

Item	Postgraduate vs. Bachelor	Postgraduate vs. Diploma	Postgraduate vs. Other	Bachelor vs. Diploma	Bachelor vs. Other	Diploma vs. Other
When disagreements occur, we work hard to achieve "win-win" solutions.	0.560	0.472	0.723	0.704	0.984	0.906
It is easy to reach consensus, even on difficult issues.	0.442	0.313	0.177	0.533	0.277	0.415
There is a clear agreement about the right way and the wrong way to do things.	0.994	0.907	0.885	0.824	0.866	0.936
We often have trouble reaching agreement on key issues.	0.133	0.027	0.851	0.090	0.411	0.162

Influence of Job Duty

All three job duty groups held positive views of their school's agreement (see Table 5.17). Participants from different job duty groups differed in their perceptions of the hard working to achieve "win-win" solutions when disagreements occur and the trouble of reaching agreement on key issues. However, there was no significant difference in how the three duty groups viewed the success rate of reaching consensus and the existence of clear agreement about the right and wrong way to do things.

The *P*-values of post-hoc tests between job duty groups are shown in Table 5.18. With regard to the hard working to achieving "win-win" solutions when disagreements occur, middle leadership reported higher values than teachers, who in turn reported higher values than principals. Concerning the trouble of reaching agreement on key issues, teachers reported lower values than both middle leadership and principals.

Influence of Job Title

All three job title groups had positive views of their school's agreement (see Table 5.19). Nonetheless, the groups showed statistically significant differences in their responses to all four items.

Table 5.17 Participants' perceptions of agreement by job duty

Item	Job duty	N	M	SD	F	P
When disagreements occur, we work hard to achieve "win-win" solutions.	Teacher	1714	4.24	0.95	7.27	0.001
	Middle leadership	125	4.49	0.74		
	Principal	38	3.87	0.96		
It is easy to reach consensus, even on difficult issues.	Teacher	1710	4.33	0.85	1.35	0.259
	Middle leadership	125	4.42	0.75		
	Principal	38	4.18	0.80		
There is a clear agreement about the right way and the wrong way to do things.	Teacher	1712	4.24	0.92	1.60	0.203
	Middle leadership	125	4.35	0.81		
	Principal	38	4.42	0.83		
We often have trouble reaching agreement on key issues.	Teacher	1711	4.24	1.19	7.36	0.001
	Middle leadership	124	4.48	1.03		
	Principal	38	4.84	0.37		

Table 5.18 Participants' perceptions of agreement by job duty: Post-hoc test

Item	Teacher vs. Middle leadership	Teacher vs. Principal	Middle leadership vs. Principal
When disagreements occur, we work hard to achieve "win-win" solutions.	0.004	0.016	<0.001
It is easy to reach consensus, even on difficult issues.	0.219	0.297	0.124
There is a clear agreement about the right way and the wrong way to do things.	0.178	0.221	0.682
We often have trouble reaching agreement on key issues.	0.022	0.002	0.098

Table 5.19 Participants' perceptions of agreement by job title

Item	Job title	N	M	SD	F	P
When disagreements occur, we work hard to achieve "win-win" solutions.	Junior	423	4.30	0.90	11.16	<0.001
	Middle	466	4.07	0.95		
	Senior	999	4.31	0.95		
It is easy to reach consensus, even on difficult issues.	Junior	422	4.29	0.83	10.69	<0.001
	Middle	466	4.20	0.85		
	Senior	997	4.41	0.84		
There is a clear agreement about the right way and the wrong way to do things.	Junior	422	4.20	0.93	15.00	<0.001
	Middle	464	4.07	0.95		
	Senior	998	4.35	0.89		
We often have trouble reaching agreement on key issues.	Junior	421	4.21	1.20	3.38	0.034
	Middle	466	4.17	1.12		
	Senior	997	4.33	1.17		

The P-values of post-hoc tests between all possible combinations of job title groups are shown in Table 5.20. Concerning the hard working to achieving "win-win" solutions when disagreements occur, the Junior and Senior groups reported higher values than the Middle group. Regarding the rate of success in reaching consensus, both Junior and Middle groups reported lower values than the Senior group. In terms of the existence of clear agreement about the right and wrong way to do things, the Senior group reported higher values than both the Junior and Middle groups. Regarding the trouble of reaching agreement on key issues, the Senior group reported higher values than the Middle group.

Table 5.20 Participants' perceptions of agreement by job title: Post-hoc test

Item	Junior vs. Middle	Junior vs. Senior	Middle vs. Senior
When disagreements occur, we work hard to achieve "win-win" solutions.	<0.001	0.800	<0.001
It is easy to reach consensus, even on difficult issues.	0.106	0.014	<0.001
There is a clear agreement about the right way and the wrong way to do things.	0.031	0.007	<0.001
We often have trouble reaching agreement on key issues.	0.610	0.087	0.017

Table 5.21 Participants' perceptions of agreement by years in teaching

Item	Years in Teaching	N	M	SD	F	P
When disagreements occur, we work hard to achieve "win-win" solutions.	<6	380	4.27	0.89	0.38	0.683
	6–15	639	4.22	0.93		
	>15	931	4.23	0.98		
It is easy to reach consensus, even on difficult issues.	<6	380	4.21	0.83	4.45	0.012
	6–15	637	4.34	0.81		
	>15	928	4.36	0.87		
There is a clear agreement about the right way and the wrong way to do things.	<6	381	4.18	0.89	1.10	0.333
	6–15	635	4.25	0.90		
	>15	929	4.26	0.95		
We often have trouble reaching agreement on key issues.	<6	380	4.13	1.22	2.81	0.060
	6–15	635	4.28	1.16		
	>15	929	4.29	1.16		

Influence of Number of Years in Teaching

All three groups, those with less than 6 years of teaching experience, 6–15 years of experience, and more than 15 years of teaching experience, held positive views of their school's agreement (see Table 5.21). There were no significant differences among the three groups with different amounts of experience on the hard work of achieving "win-win" solutions when disagreements occur, the existence of clear agreement about the right and wrong way to do things, or the trouble of reaching agreement on key issues. Groups with different amounts of experience differed in their perceptions of the rate of success of reaching consensus.

The *P*-values of post-hoc tests between all possible combinations of years in teaching groups are shown in Table 5.22. Concerning the rate of success of reaching consensus, the group with less than 6 years of teaching experience, reported lower values than the groups with 6–15 years of experience or more than 15 years of teaching experience.

Influence of the School's Geographical Location

All four school geographical location groups felt positively about their school's agreement (see Table 5.23). However, there were significant differences among the four school geographical location groups on all four items being evaluated.

The *P*-values of post-hoc tests between all possible combinations of school geographical location groups are shown in Table 5.24. Concerning the work required to achieve "win-win" solutions when disagreements occur, participants from District A reported lower values than participants from the other three districts. Participants from District B reported higher values than participants from Districts C and D. Regarding the rate of success in reaching consensus, participants from District A reported lower values than participants from the other three districts. Participants from District B reported higher values than participants from District D. Concerning the existence of clear agreement about the right and wrong way to do things, participants from District A reported lower values than participants from the other three districts. Participants from District B reported higher values than participants from District C and District D. Regarding the trouble of reaching agreement on key issues, participants from District A reported lower values than participants from the other three districts. Participants from District B reported higher values than participants from District C and District D. There were no differences between participants from District C and District D on these four items.

Table 5.22 Participants' perceptions of agreement by years in teaching: Post-hoc test

Item	<6 vs. 6–15	<6 vs. >15	6–15 vs. >15
When disagreements occur, we work hard to achieve "win-win" solutions.	0.402	0.458	0.858
It is easy to reach consensus, even on difficult issues.	0.014	0.004	0.739
There is a clear agreement about the right way and the wrong way to do things.	0.202	0.157	0.947
We often have trouble reaching agreement on key issues.	0.048	0.022	0.829

Table 5.23 Participants' perceptions of agreement by school location

Item	Location	N	M	SD	F	P
When disagreements occur, we work hard to achieve "win-win" solutions.	District A	355	3.59	1.05	77.58	<0.001
	District B	1011	4.42	0.85		
	District C	133	4.15	1.02		
	District D	486	4.32	0.84		
It is easy to reach consensus, even on difficult issues.	District A	356	3.74	0.94	81.84	<0.001
	District B	1008	4.50	0.74		
	District C	133	4.45	0.79		
	District D	484	4.31	0.82		
There is a clear agreement about the right way and the wrong way to do things.	District A	354	3.59	1.06	83.13	<0.001
	District B	1009	4.44	0.81		
	District C	133	4.25	0.85		
	District D	485	4.26	0.86		
We often have trouble reaching agreement on key issues.	District A	356	3.86	1.12	26.99	<0.001
	District B	1007	4.46	1.10		
	District C	133	4.18	1.25		
	District D	484	4.14	1.22		

Table 5.24 Participants' perceptions of agreement by location: Post-hoc test

Item	District A vs. B	District A vs. C	District A vs. D	District B vs. C	District B vs. D	District C vs. D
When disagreements occur, we work hard to achieve "win-win" solutions.	<0.001	<0.001	<0.001	0.001	0.040	0.053
It is easy to reach consensus, even on difficult issues.	<0.001	<0.001	<0.001	0.467	<0.001	0.077
There is a clear agreement about the right way and the wrong way to do things.	<0.001	<0.001	<0.001	0.016	<0.001	0.891
We often have trouble reaching agreement on key issues.	<0.001	0.006	<0.001	0.007	<0.001	0.736

5.3 Coordination and Integration

In order to understand participants' perceptions of the coordination and integration of their school culture, four items—"People from different parts of the school share a common perspective;" "It is easy to coordinate work across different parts of the school;" "It is difficult for teachers to cooperate at our school;" "There is good alignment of goals across all levels of the school"—were used that required five-point Likert-type responses:

Overall Findings

Table 5.25 shows a summary of the descriptive data of participants' responses to these items. In general, participants evaluated higher values than the mean on all four items, especially on the last item. Participants generally held positive views of their school's coordination and integration.

Influence of Gender

Both male and female participants held positive views of their school's coordination and integration (see Table 5.26). Females were more likely than males to feel that there was a common perspective among people from different parts of the school. Females were less likely to feel that cooperation was difficult in their school, more likely to feel that goals were aligned across all levels, than males. Males and females had similar opinions on the ease of coordinating work across different parts of the school.

Table 5.25 Participants' perceptions of coordination and integration

Item	N	M	SD
People from different parts of the school share a common perspective.	1981	4.30	0.90
It is easy to coordinate working across different parts of the school.	1980	4.13	0.95
It is difficult for teachers to cooperate at our school.	1977	4.30	1.11
There is good alignment of goals across all levels.	1976	4.37	0.86

Table 5.26 Participants' perceptions of coordination and integration by gender

Item	Gender	N	M	SD	t	P
People from different parts of the school share a common perspective.	Male	353	4.17	0.96	−3.05	0.002
	Female	1613	4.33	0.89		
It is easy to coordinate work across different parts of the school.	Male	351	4.06	1.00	−1.51	0.132
	Female	1613	4.14	0.94		
It is difficult for teachers to cooperate at our school.	Male	353	4.07	1.29	−4.38	<0.001
	Female	1608	4.35	1.07		
There is good alignment of goals across all levels.	Male	350	4.25	0.95	−2.90	0.004
	Female	1611	4.40	0.84		

Table 5.27 Participants' perceptions of coordination and integration by educational level

Item	Education	N	M	SD	F	P
People from different parts of the school share a common perspective.	Postgraduate	74	4.28	0.80	1.04	0.373
	Bachelor's degree	1529	4.28	0.92		
	Diploma	282	4.38	0.86		
	Other	35	4.23	0.88		
It is easy to coordinate work across different parts of the school.	Postgraduate	74	4.09	0.88	0.67	0.573
	Bachelor's degree	1528	4.11	0.96		
	Diploma	281	4.20	0.95		
	Other	35	4.17	0.95		
It is difficult for teachers to cooperate at our school.	Postgraduate	74	3.96	1.32	3.64	0.012
	Bachelor's degree	1526	4.30	1.11		
	Diploma	281	4.42	1.05		
	Other	34	4.12	1.27		
There is good alignment of goals across all levels.	Postgraduate	74	4.30	0.84	2.25	0.081
	Bachelor's degree	1525	4.35	0.87		
	Diploma	281	4.49	0.81		
	Other	35	4.43	0.85		

Influence of Educational Level

All four educational level groups held positive views of their school's coordination and integration (see Table 5.27). There were no significant differences among educational level groups on the existence of a common perspective among people from different parts of the school, the ease of coordinating work across different parts in the school, and the alignment of goals across all levels. Participants from different educational level groups differed in their perceptions of the difficulty of cooperation at their school.

The P-values of post-hoc tests between all possible combinations of educational level groups are shown in Table 5.28. With regard to teachers having difficulties cooperating at their school, participants who had postgraduate degrees reported lower values than participants with bachelor's degrees and diplomas.

Influence of Job Duty

All three job duty groups had positive views of their school's coordination and integration (see Table 5.29). Participants from different job duty groups had similar perceptions of the existence of a common perspective among people from different parts of the school, the ease of coordinating work across different parts of the school, and the alignment of goals across all levels.

Table 5.28 Participants' perceptions of coordination and integration by educational level: Post-hoc test

Item	Postgraduate vs. Bachelor	Postgraduate vs. Diploma	Postgraduate vs. Other	Bachelor vs. Diploma	Bachelor vs. Other	Diploma vs. Other
People from different parts of the school share a common perspective.	0.996	0.401	0.766	0.089	0.724	0.340
It is easy to coordinate work across different parts of the school.	0.879	0.416	0.694	0.175	0.715	0.887
It is difficult for teachers to cooperate at our school.	0.010	0.002	0.493	0.104	0.348	0.139
There is good alignment of goals across all levels.	0.573	0.084	0.455	0.014	0.614	0.684

However, there were differences in opinion among job duty groups regarding the difficulty of cooperation at their school.

The *P*-values of post-hoc tests between all possible combinations of job duty groups are shown in Table 5.30. Regarding the perceptions of the difficulty of cooperation at their school, teachers reported lower values than middle leadership.

Influence of Job Title

All three job title groups held positive perceptions of their school's coordination and integration (see Table 5.31). There were significant differences among the three job title groups on all four items.

The *P*-values of post-hoc tests between all possible combinations of job title groups are shown in Table 5.32. Regarding the existence of a common perspective among people from different parts of the school, the Junior and Middle groups reported lower values than the Senior group. With regard to the coordinating work across different parts of the school, the Senior group reported higher values than the Junior group, which in turn reported higher values than the Middle group. Concerning the challenges of cooperation at

Table 5.29 Participants' perceptions of coordination and integration by job duty

Item	Job duty	N	M	SD	F	P
People from different parts of the school share a common perspective.	Teacher	1712	4.31	0.91	0.16	0.848
	Middle leadership	124	4.35	0.84		
	Principal	38	4.29	0.77		
It is easy to coordinate work across different parts of the school.	Teacher	1710	4.13	0.96	1.55	0.213
	Middle leadership	125	4.23	0.87		
	Principal	38	4.34	0.78		
It is difficult for teachers to cooperate at our school.	Teacher	1709	4.28	1.13	6.89	0.001
	Middle leadership	125	4.63	0.76		
	Principal	38	4.58	0.76		
There is good alignment of goals across all levels.	Teacher	1708	4.38	0.87	2.12	0.120
	Middle leadership	125	4.48	0.71		
	Principal	38	4.16	0.75		

Table 5.30 Participants' perceptions of coordination and integration by job duty: Post-hoc test

Item	Teacher vs. Middle leadership	Teacher vs. Principal	Middle leadership vs. Principal
People from different parts of the school share a common perspective.	0.579	0.898	0.696
It is easy to coordinate work across different parts of the school.	0.245	0.173	0.531
It is difficult for teachers to cooperate at our school.	0.001	0.104	0.795
There is good alignment of goals across all levels.	0.203	0.117	0.043

their school, the Senior group reported higher values than both the Junior and Middle groups. Regarding the alignment of goals across all levels, both the Junior and Senior groups reported higher values than the Middle group.

Influence of Number of Years in Teaching

All three groups, those with less than 6 years of teaching experience, 6–15 years of experience, and more than 15 years of teaching experience, had positive views of their school's coordination and integration (see Table 5.33). There were no significant differences among three groups with different levels of experience on the existence of a common perspective among people

Table 5.31 Participants' perceptions of coordination and integration by job title

Item	Job title	N	M	SD	F	P
People from different parts of the school share a common perspective.	Junior	421	4.27	0.90	10.88	<0.001
	Middle	467	4.16	0.94		
	Senior	997	4.39	0.89		
It is easy to coordinate work across different parts of the school.	Junior	422	4.11	0.94	16.11	<0.001
	Middle	464	3.96	0.95		
	Senior	997	4.25	0.93		
It is difficult for teachers to cooperate at our school.	Junior	423	4.24	1.19	6.41	0.002
	Middle	464	4.19	1.11		
	Senior	995	4.40	1.07		
There is good alignment of goals across all levels.	Junior	421	4.37	0.83	10.53	<0.001
	Middle	464	4.23	0.88		
	Senior	995	4.45	0.86		

Table 5.32 Participants' perceptions of coordination and integration by job title: Post-hoc test

Item	Junior vs. Middle	Junior vs. Senior	Middle vs. Senior
People from different parts of the school share a common perspective.	0.064	0.023	<0.001
It is easy to coordinate work across different parts of the school.	0.013	0.011	<0.001
It is difficult for teachers to cooperate at our school.	0.527	0.016	0.001
There is good alignment of goals across all levels.	0.011	0.132	<0.001

from different parts of the school, the coordination of work across different parts of the school, or the alignment of goals across all levels. Groups with different levels of experience differed in their perceptions of the challenges of cooperation at their school.

The P-values of post-hoc tests between all possible combinations of years in teaching groups are shown in Table 5.34. With regard to the challenges of cooperation at their school, the group with less than six year of experience reported lower values than the groups with 6–15 years of experience or the group with more than 15 years of experience.

Influence of the School's Geographical Location

All four school geographical location groups held positive views of their school's coordination and integration (see Table 5.35). However, there were significant differences among the four groups on all four items being evaluated.

The P-values of post-hoc tests between all possible combinations of school geographical location groups are shown in Table 5.36. With regard to the

Table 5.33 Participants' perceptions of coordination and integration by years in teaching

Item	Years in Teaching	N	M	SD	F	P
People from different parts of the school share a common perspective.	<6	380	4.24	0.93	1.67	0.189
	6–15	638	4.29	0.89		
	>15	928	4.34	0.90		
It is easy to coordinate work across different parts of the school.	<6	380	4.03	0.95	2.55	0.078
	6–15	636	4.15	0.93		
	>15	928	4.16	0.97		
It is difficult for teachers to cooperate atour school.	<6	380	4.09	1.25	8.11	<0.001
	6–15	634	4.34	1.10		
	>15	927	4.36	1.06		
There is good alignment of goals across all levels.	<6	379	4.32	0.84	1.08	0.338
	6–15	635	4.37	0.85		
	>15	927	4.40	0.88		

existence of a common perspective among people from different parts of the school, participants from District A reported lower values than participants from the other three districts. Participants from District B reported higher values than participants from Districts C and D. With regard to the coordination of work across different parts of the school, participants from District A reported lower values than participants from the other three districts. Participants from District B reported higher values than participants from District D, which in turn reported higher values than participants from District C. Concerning the challenges of cooperation at their

Table 5.34 Participants' perceptions of coordination and integration by years in teaching: Post-hoc test

Item	<6 vs. 6–15	<6 vs. >15	6–15 vs. >15
People from different parts of the school share a common perspective.	0.373	0.075	0.323
It is easy to coordinate work across different parts of the school.	0.065	0.028	0.777
It is difficult for teachers to cooperate at our school.	0.001	<0.001	0.775
There is good alignment of goals across all levels.	0.466	0.150	0.433

Table 5.35 Participants' perceptions of coordination and integration by school location

Item	Location	N	M	SD	F	P
People from different parts of the school share a common perspective.	District A	356	3.73	1.03	70.30	<0.001
	District B	1007	4.50	0.79		
	District C	133	4.29	0.86		
	District D	485	4.30	0.87		
It is easy to coordinate work across different parts of the school.	District A	356	3.51	1.06	81.54	<0.001
	District B	1009	4.35	0.81		
	District C	132	3.89	1.08		
	District D	483	4.18	0.89		
It is difficult for teachers to cooperate at our school.	District A	355	3.87	1.08	35.55	<0.001
	District B	1009	4.52	1.01		
	District C	131	4.30	1.07		
	District D	482	4.16	1.24		
There is good alignment of goals across all levels.	District A	355	3.74	1.01	96.97	<0.001
	District B	1006	4.59	0.70		
	District C	133	4.32	0.92		
	District D	482	4.39	0.81		

Table 5.36 Participants' perceptions of coordination and integration by location: Post-hoc test

Item	District A vs. B	District A vs. C	District A vs. D	District B vs. C	District B vs. D	District C vs. D
People from different parts of the school share a common perspective.	<0.001	<0.001	<0.001	0.008	<0.001	0.907
It is easy to coordinate work across different parts of the school.	<0.001	<0.001	<0.001	<0.001	0.001	0.001
It is difficult for teachers to cooperate at our school.	<0.001	<0.001	<0.001	0.025	<0.001	0.184
There is good alignment of goals across all levels.	<0.001	<0.001	<0.001	<0.001	<0.001	0.414

school, participants from District A reported lower values than participants from the other three districts. Participants from District B reported higher values than participants from Districts C and D. Regarding the alignment of goals across all levels, participants from District A reported lower values than participants from the other three districts. Participants from District B reported higher values than participants from Districts C and D.

5.4 Summary

The purpose of this chapter was to investigate school staff members' opinions of consistency, one of the school culture traits. There have been three sections in this chapter, one for each of the three indexes of consistency: core values, team orientation, and coordination and integration. Four items were used for each area, using a five-point Likert-type item, which the participants then responded to. First, participants' ideas on core values were presented. Four items are used: "I am aware of the principal's educational philosophy and can adjust my own accordingly;" "I understand the school's development goals;" "I understand the school motto's cultural meaning;" "My mode of dress and behavior are consistent with school culture." Second, participants' ideas on agreement are presented. Four items are used: "When disagreements occur, we work hard to achieve 'win-win' solutions;" "It is easy to reach consensus, even on difficult issues;" "There is a clear agreement about the right way and the wrong way to do things;" "We often have trouble reaching agreement on key issues." The third section reports participants' assessments of capability development. There are four items in this section: "People from different parts of the school share a common perspective;" "It is easy to coordinate work across different parts of the school;" "It is difficult for teachers to cooperate at our school;" "There is good alignment of goals across all levels."

All twelve items are five-point Likert-type questions. The overall findings based on these questions were reported in this chapter. The effects of variables such as gender, educational level, job duty, job title, years of teaching, and school location were considered for each item.

The overall findings of the participants' responses to the items were recorded. The effects of variables such as gender, educational level, job duty, job title, years in teaching, and school location were considered for each item. The findings presented in this chapter can be summarized as follows:

1. Participants in this investigation generally agreed on the core values of their school culture. They were aware of the principal's educational philosophy and could adjust their own accordingly. They could understand the school's development goals. They could also understand the school motto's cultural meaning. They thought their mode of dress and behavior were consistent with school culture. However, significant differences can be observed between some groups; for example, Middle leadership reported lower values than both teachers and principals, and the Senior group reported a higher degree of agreement than the Junior group, which reported higher than Middle group in turn, and participants from District

A reported a lower degree of agreement than participants from the other three districts. The factor of educational level did not have any effect on this index.

2. Findings from agreement showed that staff members overall held positive views of the school culture index at their school. Participants worked hard to achieve "win-win" solutions when disagreements occurred. They usually reached consensus, even on difficult issues. They thought there was clear agreement at their school about the right and wrong way to do things. They did not have trouble reaching agreements on key issues. The factor of educational level did not influence agreement. However, differences can be found among job title groups and school location groups. Concerning the effects of job title, the Middle group reported a lower degree of agreement than the Senior group. Participants from District A reported a lower degree of agreement than participants from the other three districts, and participants from Districts C and D reported similar degrees of agreement.
3. Generally, participants agreed on the coordination and integration of their school's culture. They thought teachers from different parts of the school shared a common perspective. They agreed that it was easy to coordinate work across different parts of the school. They also agreed that it was easy for teachers to cooperate at their school. They thought there was an alignment of goals across all levels. Nevertheless, significant differences can be observed among some groups. The Middle group reported a lower degree of agreement than the Senior group. There was a significant difference among school location groups; that is, participants from District A reported a lower degree of agreement than participants from the other three districts, and participants from District B reported higher values than participants from District D.

In conclusion, participants strongly agreed that their schools have good school cultures in terms of consistency. The extent of agreement differed among groups based on gender and job duty. Furthermore, school location and staff members' job title played an important role in their opinions on their school's consistency. The results also indicate further need to enhance consistency for participants from District A and participants in the Middle job title group.

6

Adaptability

This chapter explores school staff members 'perceptions of one of the school culture traits: *adaptability*. Adaptability consists of three indexes, namely *creating change, customer focus,* and *organizational learning*. Twelve survey items were used as a way to evaluate these indexes. The findings are presented and analyzed in three sections. Section one reports how participants responded to questions about the flexibility in both teaching and management modes, effective strategies for competing with peer schools, new and improved ways to carry out work, resistance to attempts to create change. Section two presents participants' perceptions of changes made due to parents' comments and recommendations, teachers' understanding of the parents' wants and needs, taking parents' interests into consideration, and encouraging teachers to be in direct contact with parents. Section three shows how participants responded to questions about failure as an opportunity for learning and improvement, the encouragement and reward of innovation and risk taking, leaving questions unsettled, and the importance of learning in their day-to-day work. Each section details the overall findings for the given items as well as the effects of participants' gender, educational level, job duty, job title, years in teaching, and school location on the findings.

6.1 Creating Change

In order to understand participants' perceptions of creating change in their school culture, four items— "The mode of both teaching and management is very flexible;" "The school has an effective strategy for competing with peer schools;" "New and improved ways to do work are continually adopted;" "Attempts to create change usually meet with resistance" —were used that required five-point Likert-type responses.

Overall Findings

A summary of the descriptive data of the participants' responses to these items is shown in Table 6.1. In general, participants reported higher values than the

Table 6.1 Participants' perceptions of creating change

Item	N	M	SD
The mode of both teaching and management is very flexible.	1948	4.00	0.97
The school has an effective strategy for competing with peer schools.	1980	4.25	0.92
New and improved ways to do work are continually adopted.	1944	4.24	0.90
Attempts to create change usually meet with resistance.	1940	3.73	1.22

mean on all items, especially on the second and third ones. Participants took a favorable view of their school's ability to create change.

Influence of Gender

Both male and female participants felt positively about their school's performance in creating change (see Table 6.2). Males and females had similar perceptions of the flexibility of both teaching and management mode, effective strategies for competing with peer schools, and the new and improved ways to carry out work. In general, females reported higher values on resistance to attempts to create change than males.

Influence of Educational Level

All four educational level groups held positive opinions of their school's performance in creating change (see Table 6.3). There were no significant differences among the four educational level groups on the flexibility of both teaching and management mode, effective strategies for competing with peer schools, and the new and improved ways to carry out work. Participants from different educational level groups differed in their perceptions of resistance to attempts to create change.

Table 6.2 Participants' perceptions of creating change by gender

Item	Gender	N	M	SD	t	P
The mode of both teaching and management is very flexible.	Male	351	3.96	0.96	–0.91	0.365
	Female	1582	4.01	0.97		
The school has an effective strategy for competing with peer schools.	Male	353	4.26	0.86	0.06	0.949
	Female	1612	4.25	0.92		
New and improved ways to do work are continually adopted.	Male	349	4.17	0.87	–1.71	0.087
	Female	1580	4.26	0.90		
Attempts to create change usually meet with resistance.	Male	347	3.35	1.29	–6.40	<0.001
	Female	1578	3.81	1.19		

Table 6.3 Participants' perceptions of creating change by educational level

Item	Job title	N	M	SD	F	P
The mode of both teaching and management is very flexible.	Postgraduate degree	74	4.01	0.91	0.61	0.606
	Bachelor's degree	1502	3.99	0.99		
	Diploma	274	4.07	0.89		
	Other	36	4.03	1.00		
The school has an effective strategy for competing with peer schools.	Postgraduate degree	74	4.30	0.79	0.42	0.740
	Bachelor's degree	1527	4.25	0.91		
	Diploma	281	4.30	0.88		
	Other	36	4.19	1.04		
New and improved ways to do work are continually adopted.	Postgraduate degree	74	4.20	0.88	0.50	0.681
	Bachelor's degree	1499	4.24	0.89		
	Diploma	274	4.29	0.89		
	Other	36	4.14	1.05		
Attempts to create change usually meet with resistance.	Postgraduate degree	74	3.00	1.25	9.19	<0.001
	Bachelor's degree	1498	3.75	1.21		
	Diploma	271	3.77	1.22		
	Other	35	3.69	1.35		

The P-values of post-hoc tests between all possible combinations of educational level groups are shown in Table 6.4. With regard to resistance to change, participants with postgraduate education reported lower values than the other three educational groups.

Influence of Job Duty

All three job duty groups held positive perceptions of their school's performance in the area of creating change (see Table 6.5). Teachers, middle leadership, and principals had similar perceptions of the flexibility of teaching, new and improved ways to carry out work, and resistance to change. However, the groups differed in their perceptions of effective strategies for competing with peer schools.

The P-values of post-hoc tests between all possible combinations of job duty groups are shown in Table 6.6. Middle leadership reported higher values than teachers, which in turn reported higher than principal on effective strategies for competing with peer schools.

Influence of Job Title

All three job title groups had favorable opinions of their school's performance in creating change (see Table 6.7), though there were significant differences among the three groups on all four items.

Table 6.4 Participants' perceptions of creating change by educational level: Post-hoc test

Item	Postgraduate vs. Bachelor	Postgraduate vs. Diploma	Postgraduate vs. Other	Bachelor vs. Diploma	Bachelor vs. Other	Diploma vs. Other
The mode of both teaching and management is very flexible.	0.821	0.640	0.942	0.180	0.805	0.793
The school has an effective strategy for competing with peer schools.	0.632	0.965	0.577	0.334	0.738	0.501
New and improved ways to do work are continually adopted.	0.733	0.444	0.724	0.363	0.505	0.332
Attempts to create change usually meet with resistance.	<0.001	<0.001	0.006	0.752	0.758	0.682

Table 6.5 Participants' perceptions of creating change by job duty

Item	Job duty	N	M	SD	F	P
The mode of both teaching and management is very flexible.	Teacher	1680	4.00	0.98	1.65	0.192
	Middle leadership	124	4.16	0.83		
	Principal	38	3.95	0.84		
The school has an effective strategy for competing with peer schools.	Teacher	1710	4.26	0.92	3.49	0.031
	Middle leadership	125	4.44	0.71		
	Principal	38	4.05	0.80		
New and improved ways to do work are continually adopted.	Teacher	1676	4.25	0.89	2.39	0.092
	Middle leadership	124	4.39	0.76		
	Principal	38	4.05	0.84		
Attempts to create change usually meet with resistance.	Teacher	1673	3.74	1.22	0.63	0.534
	Middle leadership	123	3.80	1.21		
	Principal	38	3.55	1.06		

The P-values of post-hoc tests between all possible combinations of job title groups are shown in Table 6.8. On the flexibility of both teaching and management mode, Junior and Middle groups reported lower values than the

Table 6.6 Participants' perceptions of creating change by job duty: Post-hoc test

Item	Teacher vs. Middle leadership	Teacher vs. Principal	Middle leadership vs. Principal
The mode of both teaching and management is very flexible.	0.077	0.732	0.233
The school has an effective strategy for competing with peer schools.	0.028	0.171	0.021
New and improved ways to do work are continually adopted.	0.101	0.168	0.041
Attempts to create change usually meet with resistance.	0.561	0.351	0.264

Table 6.7 Participants' perceptions of creating change by job title

Item	Job title	N	M	SD	F	P
The mode of both teaching and management is very flexible.	Junior	417	3.93	1.01	15.08	<0.001
	Middle	456	3.84	1.00		
	Senior	978	4.12	0.93		
The school has an effective strategy for competing with peer schools.	Junior	423	4.22	0.90	11.66	<0.001
	Middle	463	4.12	0.96		
	Senior	997	4.36	0.88		
New and improved ways to do work are continually adopted.	Junior	414	4.22	0.90	13.21	<0.001
	Middle	455	4.10	0.93		
	Senior	978	4.35	0.84		
Attempts to create change usually meet with resistance.	Junior	413	3.79	1.17	25.78	<0.001
	Middle	455	3.40	1.21		
	Senior	975	3.88	1.21		

Table 6.8 Participants' perceptions of creating change by job title: Post-hoc test

Item	Junior vs. Middle	Junior vs. Senior	Middle vs. Senior
The mode of both teaching and management is very flexible.	0.199	0.001	<0.001
The school has an effective strategy for competing with peer schools.	0.113	0.008	<0.001
New and improved ways to do work are continually adopted.	0.046	0.011	<0.001
Attempts to create change usually meet with resistance.	<0.001	0.182	<0.001

Senior group. With regard to effective strategies for competing with peer schools, Junior and Middle groups reported lower values than the Senior group. Concerning new and improved ways to carry out work, the Senior group reported higher values than the Junior group, which in turn reported higher values than the Middle group. Regarding resistance to attempts to create change, Junior and Senior groups reported higher values than the Middle group.

Influence of Number of Years in Teaching

All three of the teaching groups, those with less than 6 years of teaching experience, 6–15 years of experience, and more than 15 years of teaching experience, held positive opinions of their school's performance in the area of creating change (see Table 6.9). There were no significant differences among the three groups on the first three items. Groups with different amounts of experience differed in their perceptions of the resistance to create change.

The P-values of post-hoc tests between all possible combinations of the groups according to number of years in teaching are shown in Table 6.10. Regarding resistance to attempts to create change, the group with less than 6 years of experience reported lower values than the group with more than 15 years of experience.

Influence of the School's Geographical Location

All four school geographical location groups had positive perceptions of their school's performance in the area of creating change (see Table 6.11). However,

Table 6.9 Participants' perceptions of creating change by number of years in teaching

Item	Years in Teaching	N	M	SD	F	P
The mode of both teaching and management is very flexible.	<6	376	3.97	0.97	0.54	0.585
	6–15	628	3.99	0.97		
	>15	909	4.03	0.96		
The school has an effective strategy for competing with peer schools.	<6	381	4.21	0.88	0.74	0.477
	6–15	636	4.27	0.89		
	>15	927	4.27	0.94		
New and improved ways to do work are continually adopted.	<6	374	4.25	0.84	0.26	0.775
	6–15	626	4.23	0.90		
	>15	909	4.26	0.90		
Attempts to create change usually meet with resistance.	<6	374	3.56	1.27	5.62	0.004
	6–15	626	3.71	1.22		
	>15	905	3.81	1.20		

Table 6.10 Participants' perceptions of creating change by years inteaching: Post-hoc test

Item	<6 vs. 6–15	<6 vs. >15	6–15 vs. >15
The mode of both teaching and management is very flexible.	0.749	0.343	0.473
The school has an effective strategy for competing with peer schools.	0.321	0.236	0.878
New and improved ways to do work are continually adopted.	0.713	0.832	0.476
Attempts to create change usually meet with resistance.	0.064	0.001	0.112

Table 6.11 Participants' perceptions of creating change by school location

Item	Location	N	M	SD	F	P
The mode of both teaching and management is very flexible.	District A	357	3.54	1.01	39.65	<0.001
	District B	975	4.17	0.90		
	District C	132	3.82	1.05		
	District D	484	4.05	0.96		
The school has an effective strategy for competing with peer schools.	District A	355	3.65	1.06	75.31	<0.001
	District B	1008	4.44	0.82		
	District C	133	4.11	0.91		
	District D	484	4.34	0.80		
New and improved ways to do work are continually adopted.	District A	357	3.66	0.99	74.22	<0.001
	District B	972	4.44	0.79		
	District C	133	4.17	0.94		
	District D	482	4.29	0.83		
Attempts to create change usually meet with resistance.	District A	355	3.27	1.10	37.80	<0.001
	District B	971	4.00	1.16		
	District C	132	3.52	1.12		
	District D	482	3.59	1.29		

there were significant differences among the four school geographical location groups on all four statements being evaluated.

The P-values of post-hoc tests between all possible combinations of school geographical location groups are shown in Table 6.12. On the flexibility of both teaching and management mode, participants from District A reported lower values than participants from the other three districts, and participants from District B reported higher values than participants from District D, who in turn reported higher values than participants from District C. With regard to effective strategies for competing with peer schools, participants from District A reported lower values than participants from the other three districts, and participants from District B reported higher values than participants from

District D, who in turn reported higher values than participants from District C. Concerning new and improved ways to carry out work, participants from District A reported lower values than participants from the other three districts, and participants from District B reported higher values than participants from Districts C and D. Regarding resistance to attempts to create change, participants from District A reported lower values than participants from the other three districts, and participants from District B reported higher values than participants from Districts C and D.

6.2 Customer Focus

In order to understand the participants' perceptions of the customer focus of their school culture, four items— "Parent comments and recommendations often lead to changes;" "All teachers have a deep understanding of what parents want and need;" "The interests of parents often get ignored in our decisions;" "We encourage teachers to have direct contact with parents" —were asked that required five-point Likert-type responses.

Overall Findings

Table 6.13 shows a summary of the descriptive data of the participants' opinions on these four items. In general, participants reported higher values than the mean on all four items, reporting especially high values on the third one. Participants largely held a positive view of their school's customer focus.

Influence of Gender

Both male and female participants felt positively about their school's customer focus (see Table 6.14). Males and females had similar opinions on making changes due to parents' comments and recommendations, teachers' understanding of parents' wants and needs, and encouraging teachers to be in direct contact with parents. Females had a more positive perception of taking parents' interests into consideration than males.

Influence of Educational Level

All four educational level groups had favorable opinions of their school's customer focus (see Table 6.15). Participants from different educational level groups differed in their perceptions of making changes due to parents' comments and recommendations and taking parents' interests into consideration.There were no significant differences among the four educational

Table 6.12 Participants' perceptions of creating change by location: Post-hoc test

Item	District A vs. B	District A vs. C	District A vs. D	District B vs. C	District B vs. D	District C vs. D
The mode of both teaching and management is very flexible.	<0.001	0.004	<0.001	<0.001	0.028	0.013
The school has an effective strategy for competing with peer schools.	<0.001	<0.001	<0.001	<0.001	0.034	0.009
New and improved ways to do work are continually adopted.	<0.001	<0.001	<0.001	0.001	0.001	0.174
Attempts to create change usually meet with resistance.	<0.001	0.039	<0.001	<0.001	<0.001	0.555

Table 6.13 Participants' perceptions of customer focus

Item	N	M	SD
Parent comments and recommendations often lead to changes.	1929	3.32	1.06
All teachers have a deep understanding of what parents want and need.	1947	4.08	0.85
The interests of parents often get ignored in our decisions.	1981	4.58	0.90
We encourage teachers to have direct contact with parents.	1946	4.32	0.86

Table 6.14 Participants' perceptions of customer focus by gender

Item	Gender	N	M	SD	t	P
Parent comments and recommendations often lead to changes.	Male	345	3.40	1.04	1.45	0.148
	Female	1569	3.31	1.07		
All teachers have a deep understanding of what parents want and need.	Male	349	4.01	0.89	–1.77	0.077
	Female	1583	4.10	0.84		
The interests of parents often get ignored in our decisions.	Male	351	4.31	1.13	–6.18	<0.001
	Female	1614	4.64	0.84		
We encourage teachers to have direct contact with parents.	Male	351	4.28	0.88	–1.03	0.305
	Female	1580	4.33	0.85		
	Female	1611	4.17	0.99		

Table 6.15 Participants' perceptions of customer focus by educational level

Item	Job title	N	M	SD	F	P
Parent comments and recommendations often lead to changes.	Postgraduate degree	73	3.67	1.00	3.71	0.011
	Bachelor's degree	1490	3.33	1.05		
	Diploma	270	3.30	1.12		
	Other	35	3.00	1.28		
All teachers have a deep understanding of what parents want and need.	Postgraduate degree	74	4.03	0.94	0.12	0.950
	Bachelor's degree	1501	4.08	0.85		
	Diploma	275	4.07	0.85		
	Other	35	4.09	0.85		
The interests of parents often get ignored in our decisions.	Postgraduate degree	74	4.23	1.18	4.29	0.005
	Bachelor's degree	1529	4.59	0.90		
	Diploma	282	4.62	0.87		
	Other	34	4.41	0.96		
We encourage teachers to have direct contact with parents.	Postgraduate degree	74	4.32	0.80	0.21	0.892
	Bachelor's degree	1501	4.31	0.86		
	Diploma	273	4.36	0.83		
	Other	36	4.28	0.97		

level groups on teachers' understanding of what parents want and need and encouraging teachers to be in direct contact with parents.

The P-values of post-hoc tests between all possible combinations of educational level groups are shown in Table 6.16. Participants with postgraduate education reported higher values than the other three educational level groups on making changes due to parents' comments and recommendations. With regard to taking parents' interests into consideration, participants with postgraduate education reported lower values than the other three educational level groups.

Influence of Job Duty

All three job duty groups took a favorable view of their school's customer focus (see Table 6.17). There were no significant differences among the three groups with different job duties on making changes due to parents' comments and taking parents' interests into consideration. Participants from different job duty groups differed in their perceptions of teachers' understanding of what parents want and need and encouraging teachers to be in direct contact with parents.

The P-values of post-hoc tests between all possible combinations of job duty groups are shown in Table 6.18. With regard to teachers' understanding of what parents want and need, teachers and middle leadership reported higher

Table 6.16 Participants' perceptions of customer focus by educational level: Post-hoc test

Item	Postgraduate vs. Bachelor	Postgraduate vs. Diploma	Postgraduate vs. Other	Bachelor vs. Diploma	Bachelor vs. Other	Diploma vs. Other
Parent comments and recommendations often lead to changes.	0.007	0.009	0.002	0.756	0.073	0.111
All teachers have a deep understanding of what parents want and need.	0.580	0.707	0.738	0.800	0.987	0.914
The interests of parents often get ignored in our decisions.	0.001	0.001	0.334	0.525	0.267	0.198
We encourage teachers to have direct contact with parents.	0.923	0.782	0.789	0.467	0.799	0.609

Table 6.17 Participants' perceptions of customer focus by job duty

Item	Job duty	N	M	SD	F	P
Parent comments and recommendations often lead to changes.	Teacher	1664	3.32	1.07	0.39	0.675
	Middle leadership	124	3.40	1.09		
	Principal	37	3.24	0.86		
All teachers have a deep understanding of what parents want and need.	Teacher	1680	4.10	0.85	3.69	0.025
	Middle leadership	124	4.15	0.77		
	Principal	37	3.73	0.96		
The interests of parents often get ignored in our decisions.	Teacher	1712	4.58	0.90	1.76	0.172
	Middle leadership	125	4.72	0.75		
	Principal	38	4.47	0.86		
We encourage teachers to have direct contact with parents.	Teacher	1681	4.30	0.87	7.94	<0.001
	Middle leadership	123	4.59	0.65		
	Principal	38	4.58	0.60		

values than principals. Middle leadership and principals reported higher values than teachers on taking parents' interests into consideration.

Influence of Job Title

All three job title groups held positive perceptions of their school's customer focus (see Table 6.19). There were no significant differences among the three

Table 6.18 Participants' perceptions of customer focus by job duty: Post-hoc test

Item	Teacher vs. Middle leadership	Teacher vs. Principal	Middle leadership vs. Principal
Parent comments and recommendations often lead to changes.	0.450	0.663	0.446
All teachers have a deep understanding of what parents want and need.	0.464	0.010	0.008
The interests of parents often get ignored in our decisions.	0.088	0.473	0.137
We encourage teachers to have direct contact with parents.	<0.001	0.049	0.968

Table 6.19 Participants' perceptions of customer focus by job title

Item	Job title	N	M	SD	F	P
Parent comments and recommendations often lead to changes.	Junior	413	3.24	1.07	2.12	0.121
	Middle	454	3.34	1.01		
	Senior	967	3.37	1.09		
All teachers have a deep understanding of what parents want and need.	Junior	415	3.99	0.92	11.29	<0.001
	Middle	456	4.00	0.84		
	Senior	979	4.18	0.82		
The interests of parents often get ignored in our decisions.	Junior	423	4.55	0.96	6.23	0.002
	Middle	463	4.49	0.93		
	Senior	999	4.65	0.85		
We encourage teachers to have direct contact with parents.	Junior	417	4.25	0.91	9.94	<0.001
	Middle	457	4.23	0.86		
	Senior	975	4.42	0.81		

groups on changes due to parents' comments and recommendations. However, groups with different job titles differed in their perceptions of the other three items.

The P-values of post-hoc tests between all possible combinations of job title groups are shown in Table 6.20. Regarding teachers' understanding of what parents want and need, taking parents' interests into consideration, and encouraging teachers to be indirect contact with parents, the Senior group reported higher values than the Junior and Middle groups.

Influence of Number of Years in Teaching

All three groups, those with less than 6 years of teaching experience, 6–15 years of experience, and more than 15 years of teaching experience, felt positively about their school's customer focus (see Table 6.21). There were

Table 6.20 Participants' perceptions of customer focus by job title: Post-hoc test

Item	Junior vs. Middle	Junior vs. Senior	Middle vs. Senior
Parent comments and recommendations often lead to changes.	0.170	0.040	0.633
All teachers have a deep understanding of what parents want and need.	0.900	<0.001	<0.001
The interests of parents often get ignored in our decisions.	0.318	0.037	0.001
We encourage teachers to have direct contact with parents.	0.670	0.001	<0.001

Table 6.21 Participants' perceptions of customer focus by number of years in teaching

Item	Years in Teaching	N	M	SD	F	P
Parent comments and recommendations often lead to changes.	<6	374	3.34	1.06	1.23	0.293
	6–15	623	3.27	1.04		
	>15	899	3.35	1.08		
All teachers have a deep understanding of what parents want and need.	<6	375	4.01	0.90	2.32	0.098
	6–15	627	4.09	0.83		
	>15	910	4.12	0.84		
The interests of parents often get ignored in our decisions.	<6	380	4.42	1.06	8.32	<0.001
	6–15	637	4.60	0.89		
	>15	928	4.64	0.83		
We encourage teachers to have direct contact with parents.	<6	376	4.26	0.89	3.58	0.028
	6–15	627	4.29	0.86		
	>15	908	4.38	0.83		

no significant differences among the three groups with different amounts of experience on changes due to parents' comments and recommendations and teachers' understanding of what parents want and need. Groups with different amounts of experience differed in their perceptions of taking parents' interests into consideration and encouraging teachers to be indirect contact with parents.

The P-values of post-hoc tests between all possible combinations of years in teaching groups are shown in Table 6.22. With regard to taking parents' interests into consideration, the group with less than 6 years of experience reported lower values than the group with 6–15 years of experience and the group with more than 15 years. Concerning encouraging teachers to be in direct contact with parents, the group with less than 6 years of experience reported lower values than the group with more than 15 years of experience.

Influence of the School's Geographical Location

All four school geographical location groups had favorable opinions of their school's customer focus (see Table 6.23). However, there were significant differences among the four school geographical location groups on all four items being evaluated.

Table 6.22 Participants' perceptions of customer focus by number of years in teaching: Post-hoc test

Item	<6 vs. 6–15	<6 vs. >15	6–15 vs. >15
Parent comments and recommendations often lead to changes.	0.316	0.816	0.125
All teachers have a deep understanding of what parents want and need.	0.165	0.031	0.423
The interests of parents often get ignored in our decisions.	0.001	<0.001	0.455
We encourage teachers to have direct contact with parents.	0.492	0.017	0.050

Table 6.23 Participants' perceptions of customer focus by school location

Item	Location	N	M	SD	F	P
Parent comments and recommendations often lead to changes.	District A	353	3.20	0.97	4.26	0.005
	District B	968	3.32	1.11		
	District C	128	3.24	1.09		
	District D	480	3.45	1.01		
All teachers have a deep understanding of what parents want and need.	District A	354	3.74	0.86	25.30	<0.001
	District B	975	4.19	0.83		
	District C	133	4.07	0.88		
	District D	485	4.10	0.83		
The interests of parents often get ignored in our decisions.	District A	352	4.18	1.09	45.56	<0.001
	District B	1011	4.78	0.67		
	District C	133	4.59	0.83		
	District D	485	4.45	1.07		
We encourage teachers to have direct contact with parents.	District A	356	3.94	0.90	32.24	<0.001
	District B	973	4.44	0.83		
	District C	133	4.26	0.95		
	District D	484	4.37	0.76		

Table 6.24 Participants' perceptions of customer focus by location: Post-hoc test

Item	District A vs. B	District A vs. C	District A vs. D	District B vs. C	District B vs. D	District C vs. D
Parent comments and recommendations often lead to changes.	0.084	0.708	0.001	0.465	0.019	0.045
All teachers have a deep understanding of what parents want and need.	<0.001	<0.001	<0.001	0.101	0.044	0.684
The interests of parents often get ignored in our decisions.	<0.001	<0.001	<0.001	0.024	<0.001	0.092
We encourage teachers to have direct contact with parents.	<0.001	<0.001	<0.001	0.019	0.092	0.210

The P-values of post-hoc tests between all possible combinations of school geographical location groups are shown in Table 6.24. With regard to changes due to parents' comments and recommendations, participants from District D reported higher values than participants from the other three districts. With regard to teachers' understanding of what parents want and need, participants from District A reported lower values than participants from the other three districts, and participants from District B reported higher values than participants from District D. Concerning taking parents' interests into consideration, participants from District A reported lower values than participants from the other three districts, and participants from District B reported higher values than participants from Districts C and D. Regarding encouraging teachers to be in direct contact with parents, participants from District A reported lower values than participants from the other three districts, and participants from District B reported higher values than participants from District C.

6.3 Organizational Learning

In order to understand participants' perceptions of the organizational learning of their school culture, four items— "We view failure as an opportunity for learning and improvement;" "Innovation and risk taking are encouraged and

Table 6.25 Participants' perceptions of organizational learning

Item	N	M	SD
We view failure as an opportunity for learning and improvement.	1944	4.14	0.91
Innovation and risk taking are encouraged and rewarded.	1984	4.11	0.99
Many things "fall through the cracks."	1944	4.17	1.13
Learning is an important objective in our day-to-day work.	1950	4.58	0.72

rewarded;" "Manythings 'fall through the cracks;" "Learning is an important objective in our day-to-day work" —were asked that required five-point Likert-type responses.

Overall Findings

Table 6.25 shows a summary of the descriptive data of participants' responses to these items. In general, participants reported higher values than the mean on all four of these items, especially on the last one. Participants generally held a positive opinion of their school's organizational learning.

Influence of Gender

Both male and female participants took a favorable view of their school's organizational learning (see Table 6.26). Males and females had similar opinions on failure as an opportunity and risk taking being rewarded. Females agreed more strongly with statements regarding neglect of unsettled questions and the importance of learning in their day-to-day work than males.

Influence of Educational Level

All four educational level groups had positive perceptions of their school's organizational learning (see Table 6.27). There were no significant differences among the educational level groups with regard to perceiving failure as an opportunity for learning, risk taking being encouraged, and the importance of learning in their day-to-day work. Participants from different educational level groups differed in their perceptions of leaving questions unsettled (things "falling between the cracks").

The P-values of post-hoc tests between all possible combinations of educational level groups are shown in Table 6.28. Concerning leaving questions unsettled, participants with postgraduate education reported lower values than participants with a Bachelor's degree or diploma.

Table 6.26 Participants' perceptions of organizational learning by gender

Item	Gender	N	M	SD	t	P
We view failure as an opportunity for learning and improvement.	Male	348	4.11	0.90	–0.91	0.361
	Female	1581	4.16	0.91		
Innovation and risk taking	Male	353	4.16	0.94	0.76	0.448
are encouraged and rewarded.	Female	1616	4.11	0.99		
Many things "fall through	Male	350	3.97	1.26	–3.57	<0.001
the cracks."	Female	1579	4.21	1.10		
Learning is an important	Male	351	4.50	0.72	–2.34	0.019
objective in our day-to-day work.	Female	1584	4.59	0.71		

Table 6.27 Participants' perceptions of organizational learning by educational level

Item	Job title	N	M	SD	F	P
We view failure as an opportunity for learning and improvement.	Postgraduate degree	74	4.11	0.87	0.08	0.970
	Bachelor's degree	1500	4.14	0.92		
	Diploma	273	4.14	0.89		
	Other	35	4.20	0.76		
Innovation and risk taking are encouraged and rewarded.	Postgraduate degree	74	4.27	0.90	1.26	0.288
	Bachelor's degree	1531	4.13	0.98		
	Diploma	282	4.11	1.00		
	Other	36	3.89	1.06		
Many things "fall through the cracks."	Postgraduate degree	73	3.86	1.28	3.21	0.022
	Bachelor's degree	1500	4.17	1.13		
	Diploma	274	4.28	1.07		
	Other	35	3.94	1.28		
Learning is an important objective	Postgraduate degree	74	4.49	0.71	2.02	0.108
	Bachelor's degree	1503	4.57	0.73		
in our day-to-day work.	Diploma	275	4.67	0.62		
	Other	36	4.50	0.77		

Influence of Job Duty

All three job duty groups held positive perceptions of their school's organizational learning (see Table 6.29). Participants from different job duty groups differed in their perceptions of failure as an opportunity for learning and improvement, leaving questions unsettled, and the importance of learning

Table 6.28 Participants' perceptions of organizational learning by educational level: Post-hoc test

Item	Postgraduate vs. Bachelor	Postgraduate vs. Diploma	Postgraduate vs. Other	Bachelor vs. Diploma	Bachelor vs. Other	Diploma vs. Other
We view failure as an opportunity for learning and improvement.	0.750	0.771	0.623	0.997	0.713	0.727
Innovation and risk taking are encouraged and rewarded.	0.214	0.220	0.055	0.851	0.152	0.195
Many things "fall through the cracks."	0.026	0.005	0.731	0.108	0.250	0.092
Learning is an important objective in our day-to-day work.	0.336	0.055	0.926	0.038	0.571	0.191

Table 6.29 Participants' perceptions of organizational learning by job duty

Item	Job duty	N	M	SD	F	P
We view failure as an opportunity for learning and improvement.	Teacher	1677	4.15	0.92	3.89	0.021
	Middle leadership	124	4.31	0.71		
	Principal	38	3.87	0.91		
Innovation and risk taking are encouraged and rewarded.	Teacher	1714	4.12	0.98	1.56	0.210
	Middle leadership	125	4.26	0.95		
	Principal	38	3.97	0.94		
Many things "fall through the cracks."	Teacher	1677	4.16	1.14	4.30	0.014
	Middle leadership	123	4.46	0.86		
	Principal	38	4.26	0.86		
Learning is an important objective in our day-to-day work.	Teacher	1682	4.60	0.70	7.89	<0.001
	Middle leadership	124	4.65	0.59		
	Principal	38	4.16	0.86		

in their day-to-day work. However, there were no significant differences in how the three duty groups viewed rewarding innovation and risk taking.

The P-values of post-hoc tests between all possible combinations of job duty groups are shown in Table 6.30. Regarding the perception of failure as an opportunity for learning and the importance of day-to-day

Table 6.30 Participants'perceptions of organizationallearning by job duty: Post-hoc test

Item	Teacher vs. Middle leadership	Teacher vs. Principal	Middle leadership vs. Principal
We view failure as an opportunity for learning and improvement.	0.046	0.062	0.008
Innovation and risk taking are encouraged and rewarded.	0.146	0.344	0.117
Many things "fall through the cracks."	0.004	0.576	0.335
Learning is an important objective in our day-to-day work.	0.502	<0.001	<0.001

learning, middle leadership reported higher values than teachers and principals. With regard to leaving questions unsettled, teachers reported lower values than middle leadership. On the importance of learning in their day-to-day work, middle leadership reported higher values than both teachers and principals.

Influence of Job Title

All three job title groups felt positively about their school's organizational learning (see Table 6.31). None the less, the groups showed statistically significant differences in their responses on all four items being evaluated.

The P-values of post-hoc tests between all possible combinations of job title groups are shown in Table 6.32. On perceiving failure as an opportunity,

Table 6.31 Participants' perceptions of organizationallearning by job title

Item	Job title	N	M	SD	F	P
We view failure as an opportunity for learning and improvement.	Junior	415	4.12	0.93	8.96	<0.001
	Middle	456	4.02	0.91		
	Senior	978	4.23	0.89		
Innovation and risk taking are encouraged and rewarded.	Junior	423	4.17	0.95	7.56	0.001
	Middle	465	3.98	1.00		
	Senior	999	4.19	0.96		
Many things "fall through the cracks."	Junior	416	4.10	1.19	17.83	<0.001
	Middle	455	3.97	1.17		
	Senior	976	4.33	1.04		
Learning is an important objective in our day-to-day work.	Junior	417	4.58	0.68	12.76	<0.001
	Middle	456	4.45	0.79		
	Senior	980	4.65	0.68		

the Senior group reported higher values than the Junior and Middle groups. With regard to the reward of risk taking, the Middle group reported lower values than the Junior and Senior groups. Concerning leaving questions unsettled, the Senior group reported higher values than the Junior and Middle groups. Regarding the importance of learning in their day-to-day work, the Middle group reported lower values than the Junior and Senior groups.

Influence of Number of years in Teaching

All three groups, those with less than 6 years of teaching experience, 6–15 years of experience, and more than 15 years of teaching experience, had positive views of their school's organizational learning (see Table 6.33). There were no significant differences among the three groups with different levels of

Table 6.32 Participants' perceptions of organizational learning by job title: Post-hoc test

Item	Junior vs. Middle	Junior vs. Senior	Middle vs. Senior
We view failure as an opportunity for learning and improvement.	0.087	0.042	<0.001
Innovation and risk taking are encouraged and rewarded.	0.004	0.726	<0.001
Many things "fall through the cracks."	0.091	<0.001	<0.001
Learning is an important objective in our day-to-day work.	0.004	0.115	<0.001

Table 6.33 Participants' perceptions of organizational learning by years inteaching

Item	Years in Teaching	N	M	SD	F	P
We view failure as an opportunity for learning and improvement.	<6	375	4.17	0.86	0.60	0.549
	6–15	624	4.12	0.92		
	>15	910	4.16	0.92		
Innovation and risk taking are encouraged and rewarded.	<6	380	4.21	0.91	2.16	0.116
	6–15	638	4.12	0.94		
	>15	930	4.09	1.03		
Many things "fall through the cracks."	<6	376	3.96	1.29	11.25	<0.001
	6–15	625	4.14	1.15		
	>15	908	4.28	1.03		
Learning is an important objective in our day-to-day work.	<6	376	4.55	0.71	0.98	0.375
	6–15	627	4.57	0.70		
	>15	912	4.61	0.72		

Table 6.34 Participants' perceptions of organizational learning by years inteaching: Post-hoc test

Item	<6 vs. 6–15	<6 vs. >15	6–15 vs. >15
We view failure as an opportunity for learning and improvement.	0.356	0.863	0.339
Innovation and risk taking are encouraged and rewarded.	0.171	0.038	0.464
Many things "fall through the cracks."	0.017	<0.001	0.013
Learning is an important objective in our day-to-day work.	0.710	0.209	0.310

experience on failure as an opportunity for learning and improvement, the encouragement and reward of innovation and risk taking, and the importance of day-to-day learning. Groups with different levels of experience differed in their perceptions of leaving questions unsettled.

The P-values of post-hoc tests between all possible combinations of years in teaching groups are shown in Table 6.34. Concerning leaving questions unsettled, the group with more than 15 years of experience reported higher values than the group with 6–15 years of experience, which in turn reported higher values than the group with less than 6 years.

Influence of the School's Geographical Location

All four school geographical location groups had favorable opinions of their school's organizational learning (see Table 6.35). However, there were significant differences among the four groups on all four items being evaluated.

The P-values of post-hoc tests between all possible combinations of school geographical location groups are shown in Table 6.36. On perceiving failure as an opportunity for learning and improvement, participants from District A reported lower values than participants from the other three districts, and participants from District B reported higher values than participants from Districts C and D. With regard to rewarding risk taking, participants from District A reported lower values than participants from the other three districts, and participants from Districts B and D reported higher values than participants from District C. Concerning leaving questions unsettled, participants from District A reported lower values than participants from the other three districts, and participants from District B reported higher values than participants from Districts C and D. Regarding the importance of learning in their day-to-day work, participants from District A reported lower values

Table 6.35 Participants' perceptions of organizational learning by school location

Item	Location	N	M	SD	F	P
We view failure as an opportunity for learning and improvement.	District A	355	3.70	0.93	41.33	<0.001
	District B	975	4.31	0.88		
	District C	133	4.07	1.00		
	District D	481	4.15	0.84		
Innovation and risk taking are encouraged and rewarded.	District A	357	3.48	1.05	72.36	<0.001
	District B	1011	4.30	0.90		
	District C	132	3.93	1.01		
	District D	484	4.24	0.91		
Many things "fall through the cracks."	District A	355	3.65	1.20	51.93	<0.001
	District B	973	4.45	0.95		
	District C	132	4.02	1.10		
	District D	484	4.02	1.25		
Learning is an important objective in our day-to-day work.	District A	356	4.08	0.92	83.45	<0.001
	District B	976	4.73	0.57		
	District C	133	4.73	0.59		
	District D	485	4.58	0.67		

Table 6.36 Participants' perceptions of organizational learning by location: Post-hoc test

Item	District A vs. B	District A vs. C	District A vs. D	District B vs. C	District B vs. D	District C vs. D
We view failure as an opportunity for learning and improvement.	<0.001	<0.001	<0.001	0.003	0.001	0.345
Innovation and risk taking are encouraged and rewarded.	<0.001	<0.001	<0.001	<0.001	0.255	0.001
Many things "fall through the cracks."	<0.001	0.001	<0.001	<0.001	<0.001	1.000
Learning is an important objective in our day-to-day work.	<0.001	<0.001	<0.001	0.932	<0.001	0.021

than participants from the other three districts, and participants from Districts B and C reported higher values than participants from District D.

6.4 Summary

The purpose of this chapter was to investigate school staff members' perceptions of adaptability, one of the school culture traits. There have been three sections in this chapter, one for each of the three indexes of adaptability. First, participants' perceptions of their school's performance in the area of creating change were presented. Four items were used: "The mode of both teaching and management is very flexible;" "The school has an effective strategy for competing with peer schools;" "New and improved ways to do work are continually adopted;" "Attempts to create change usually meet with resistance." Second, participants' perceptions of their school's customer focus were presented. Four items were used: "Parent comments and recommendations often lead to changes;" "All teachers have a deep understanding of what parents want and need;" "The interests of parents often get ignored in our decisions;" "We encourage teachers to have direct contact with parents." Third, participants' perceptions of their school's organizational learning were presented. Four items were used: "We view failure as an opportunity for learning and improvement;" "Innovation and risk taking are encouraged and rewarded;" "Many things 'fall through the cracks;" "Learning is an important objective in our day-to-day work." All twelve items were five-point Likert-type questions. The overall findings of the participants' responses to the items were recorded. The effects of variables such as gender, educational level, job duty, job title, years in teaching, and school location were considered for each item. The findings presented in this chapter can be summarized as follows:

1. Participants in this investigation generally agreed on their school's performance in the area of creating change in their school culture. They believed the mode of teaching and management was flexible. They felt their school had an effective strategy for competing with peer schools. They agreed that there were new and improved ways to do work. They reported that attempts to create change did not meet with resistance. However, significant differences can be observed between some groups in their perceptions of creating change; for example, the Senior group reported a higher degree of agreement than the Middle group. Participants from District A reported a lower degree of agreement than participants from the other three districts.
2. Findings from the customer focus area showed a positive overall picture of staff members' agreement on this school culture index. Participants agreed that parents' comments and recommendations often lead to changes in schools. They believed all teachers had a deep understanding of parents' wants and needs. They felt that the interests of parents were not ignored in making decisions. They agreed that the school encouraged them to have direct contact with parents. However, differences can be found among school location groups. Participants from District A reported a lower degree of agreement than participants from District D.

3. Generally, participants agreed on the organizational learning of their school culture. They viewed failure as an opportunity for learning and improvement. They reported that risk taking was encouraged and rewarded. They believed that unsettled questions were not neglected. They agreed that learning was essential to their daily work. Nevertheless, there were significant differences among job title and school location groups. Concerning the effect of job title, the Middle group reported a lower degree of agreement than the Senior group. Participants from District A reported a lower degree of agreement than participants from the other three districts.

In conclusion, the participants strongly agreed that their schools have good cultures concerning adaptability. The extent of agreement differed among groups based on job duty, job title, and school location. The results also indicate further need to enhance adaptability for participants from District A and in the Middle job title group.

7

Mission

This chapter explores school staff members' perceptions of one of the school's culture traits: *mission*. Mission consists of three indexes, namely *strategic direction and intent, goals and objectives,* and *vision*. Twelve survey items were used in the survey as a way to evaluate these indexes. The findings are presented and analyzed in three sections. Section one reports how participants responded to questions about the intention of peer schools to imitate their development strategy, the consistency of applying the development strategy to school work, and the clarity of the school's future strategy and future development. Section two presents participants' perceptions of the ambitiousness and feasibility of goals set by leaders, how widely objectives were publicized, how well leaders tracked progress toward their stated goals, and the commitment of participants themselves to fulfilling the school's development goals. Section three shows how participants responded to teachers' shared vision of the school's future, the long-term viewpoint of leaders, the prevalence of compromising long-term vision with short-term thinking, and the excitement and motivation created by teachers' vision. Each section details the overall findings for the given items as well as the effects of participants' gender, educational level, job duty, job title, years in teaching, and school location, on the findings.

7.1 Strategic Direction and Intent

In order to understand participants' perceptions of the strategic direction and intent of their school culture, four items—"Peer schools in the district wish to imitate our development strategy;" "All school work is conducted with the guidance of school development strategies;" "There is a clear strategy for the future;" "Our strategic direction is unclear to me"—were used that required five-point Likert-type responses.

Overall findings

A summary of the descriptive data of the participants' responses to these items is shown in Table 7.1. In general, participants reported higher values than the mean on all items, especially on the third one. Participants held a positive perception of their school's strategic direction and intent.

Influence of Gender

Both male and female participants felt positively about their school's strategic direction and intent (see Table 7.2). Males and females had similar perceptions of the intention of peer schools to imitate their development strategy and the clarity of their future strategy. In general, females assessed the consistency with which development strategies were applied to school work and the clarity of the school's strategic direction more positively than males.

Influence of Educational Level

All four educational level groups held positive perceptions of their school's strategic direction and intent (see Table 7.3). There was no significant difference among the four educational level groups on the intention of peer schools to imitate their development strategy, the consistency with which development strategies were applied, or the clarity of future strategy. However, the groups differed in their perceptions of the clarity of their school's strategic direction.

The P-values of post-hoc tests between all possible combinations of educational level groups are shown in Table 7.4. Participants with postgraduate education reported lower values than the bachelor's and diploma educational level groups on the clarity of their school's strategic direction.

Table 7.1 Participants' perceptions of strategic direction and intent

Item	N	M	SD
Peer schools in the district wish to imitate our development strategy.	1932	3.73	1.09
All school work is conducted with the guidance of school development strategies.	1948	4.35	0.84
There is a clear strategy for the future.	1947	4.43	0.83
Our strategic direction is unclear to me.	1982	4.27	1.18

Table 7.2 Participants' perceptions of strategic direction and intent by gender

Item	Gender	N	M	SD	t	P
Peer schools in the district wish to imitate our development strategy.	Male	348	3.74	1.09	0.03	0.975
	Female	1569	3.73	1.09		
All school work is conducted with the guidance of school development strategies.	Male	351	4.27	0.88	−2.01	0.045
	Female	1582	4.37	0.83		
There is a clear strategy for the future.	Male	351	4.37	0.83	−1.63	0.103
	Female	1581	4.45	0.82		
Our strategic direction is unclear to me.	Male	353	4.04	1.31	−4.11	<0.001
	Female	1614	4.32	1.14		

Table 7.3 Participants' perceptions of strategic direction and intent by educational level

Item	Education	N	M	SD	F	P
Peer schools in the district wish to imitate our development strategy.	Postgraduate degree	73	3.96	1.02	1.10	0.348
	Bachelor's degree	1492	3.72	1.09		
	Diploma	270	3.72	1.11		
	Other	35	3.77	1.06		
All school work is conducted with the guidance of school development strategies.	Postgraduate degree	74	4.24	0.86	1.21	0.303
	Bachelor's degree	1502	4.35	0.84		
	Diploma	275	4.43	0.80		
	Other	36	4.31	0.95		
There is a clear strategy for the future.	Postgraduate degree	74	4.31	0.86	1.19	0.311
	Bachelor's degree	1501	4.43	0.83		
	Diploma	274	4.49	0.78		
	Other	36	4.33	0.93		
Our strategic direction is unclear to me.	Postgraduate degree	74	3.84	1.45	4.48	0.004
	Bachelor's degree	1528	4.27	1.19		
	Diploma	283	4.39	1.02		
	Other	36	4.11	1.19		

Influence of Job Duty

All the three job duty groups took a favorable view of their school's performance in the area of strategic direction and intent (see Table 7.5). Teachers, middle leadership, and principals had similar perceptions of peer schools' intention to imitate. However, the groups differed in their perceptions of the consistency of applying development strategies, the clarity of future strategy, and the clarity of strategic direction.

Table 7.4 Participants' perceptions of strategic direction and intent by educational level: Post-hoc test

Item	Postgraduate vs. Bachelor	Postgraduate vs. Diploma	Postgraduate vs. Other	Bachelor vs. Diploma	Bachelor vs. Other	Diploma vs. Other
Peer schools in the district wish to imitate our development strategy.	0.072	0.100	0.403	0.982	0.799	0.802
All school work is conducted with the guidance of school development strategies.	0.283	0.090	0.714	0.151	0.752	0.405
There is a clear strategy for the future.	0.225	0.094	0.893	0.252	0.487	0.278
Our strategic direction is unclear to me.	0.002	<0.001	0.254	0.107	0.437	0.184

The P-values of post-hoc tests between all possible combinations of job duty groups are shown in Table 7.6. Concerning the consistency of school work and development strategies, middle leadership reported higher values than

Table 7.5 Participants' perceptions of strategic direction and intent by job duty

Item	Job duty	N	M	SD	F	P
Peer schools in the district wish to imitate our development strategy.	Teacher	1664	3.73	1.10	0.90	0.406
	Middle leadership	124	3.85	1.05		
	Principal	38	3.61	0.89		
All school work is conducted with the guidance of school development strategies.	Teacher	1681	4.35	0.84	4.38	0.013
	Middle leadership	124	4.56	0.68		
	Principal	38	4.21	0.74		
There is a clear strategy for the future.	Teacher	1680	4.43	0.83	6.16	0.002
	Middle leadership	124	4.65	0.65		
	Principal	38	4.16	0.82		
Our strategic direction is unclear to me.	Teacher	1712	4.26	1.18	4.46	0.012
	Middle leadership	125	4.49	1.09		
	Principal	38	4.68	0.53		

Table 7.6 Participants' perceptions of strategic direction and intent by job duty: Post-hoc test

Item	Teacher vs. Middle leadership	Teacher vs. Principal	Middle leadership vs. Principal
Peer schools in the district wish to imitate our development strategy.	0.271	0.469	0.233
All school work is conducted with the guidance of school development strategies.	0.007	0.293	0.022
There is a clear strategy for the future.	0.006	0.040	0.001
Our strategic direction is unclear to me.	0.036	0.027	0.365

Table 7.7 Participants' perceptions of strategic direction and intent by job title

Item	Job title	N	M	SD	F	P
Peer schools in the district wish to imitate our development strategy.	Junior	413	3.65	1.08	16.53	<0.001
	Middle	455	3.56	1.09		
	Senior	968	3.89	1.07		
All school work is conducted with the guidance of school development strategies.	Junior	416	4.34	0.82	22.15	<0.001
	Middle	457	4.16	0.92		
	Senior	978	4.47	0.78		
There is a clear strategy for the future.	Junior	417	4.42	0.82	24.96	<0.001
	Middle	456	4.24	0.91		
	Senior	977	4.56	0.74		
Our strategic direction is unclear to me.	Junior	424	4.21	1.21	9.13	<0.001
	Middle	463	4.13	1.19		
	Senior	998	4.39	1.13		

both teachers and principals. With regard to future strategy, middle leadership reported higher values than teachers, who in turn reported higher values than principals. On the clarity of the school's strategic direction, teachers reported lower values than both middle leadership and principals.

Influence of Job Title

All three job title groups held positive opinions of their school's strategic direction and intent (see Table 7.7), though there were significant differences among the three groups on all four items.

The P-values of post-hoc tests between all possible combinations of job title groups are shown in Table 7.8. On peer schools' intention to imitate, the Senior group reported higher values than the Junior and Middle groups. With regard to consistency, the Senior group reported higher values than the Junior

Table 7.8 Participants' perceptions of strategic direction and intent by job title: Post-hoc test

Item	Junior vs. Middle	Junior vs. Senior	Middle vs. Senior
Peer schools in the district wish to imitate our development strategy.	0.241	<0.001	<0.001
All school work is conducted with the guidance of school development strategies.	0.002	0.006	<0.001
There is a clear strategy for the future.	0.001	0.002	<0.001
Our strategic direction is unclear to me.	0.333	0.006	<0.001

group, which in turn reported higher values than the Middle group. Concerning future strategy, the Senior group reported higher values than the Junior group, which in turn reported higher values than the Middle group. Regarding the clarity of the school's strategic direction, the Senior group reported higher values than the others.

Influence of Number of Years in Teaching

All three of the teaching groups, those with less than 6 years of teaching experience, 6–15 years of experience, and more than 15 years of teaching experience, had positive perceptions of their school's strategic direction and intent (see Table 7.9). There were no significant differences among the three groups on peer schools' intention to imitate or the consistency of applying development strategies in school work. Groups with different amounts of experience differed in their perceptions of the clarity of future strategy and strategic direction.

The P-values of post-hoc tests between the various groups according to the number of years in teaching are shown in Table 7.10. With regard to future strategy, the group with more than 15 years of experience reported higher values than the group with less than 6 years of experience. Concerning the clarity of the school's strategic direction, the group with less than 6 years of experience reported lower than the others.

Influence of the School's Geographical Location

All four school geographical location groups had favorable opinions of their school's strategic direction and intent (see Table 7.11). However, there were significant differences among the four school geographical location groups

Table 7.9 Participants' perception on strategic direction and intent by number of years in teaching

Item	Years in Teaching	N	M	SD	F	P
Peer schools in the district wish to imitate our development strategy.	<6	374	3.69	1.07	1.38	0.252
	6–15	624	3.71	1.08		
	>15	900	3.78	1.10		
All school work is conducted with the guidance of school development strategies.	<6	375	4.32	0.81	1.84	0.159
	6–15	628	4.32	0.86		
	>15	911	4.40	0.82		
There is a clear strategy for the future.	<6	376	4.37	0.80	3.17	0.042
	6–15	626	4.41	0.85		
	>15	910	4.48	0.81		
Our strategic direction is unclear to me.	<6	381	4.03	1.33	10.31	<0.001
	6–15	638	4.31	1.14		
	>15	927	4.34	1.13		

Table 7.10 Participants' perceptions of strategic direction and intent by years in teaching: Post-hoc test

Item	<6 vs. 6–15	<6 vs. >15	6–15 vs. >15
Peer schools in the district wish to imitate our development strategy.	0.715	0.146	0.209
All school work is conducted with the guidance of school development strategies.	0.929	0.137	0.099
There is a clear strategy for the future.	0.400	0.020	0.090
Our strategic direction is unclear to me.	<0.001	<0.001	0.536

on the intention of peer schools to imitate their development strategy, the consistency of school work and development strategies, the clarity of the school's future strategy, and the clarity of the school's strategic direction.

The P-values of post-hoc tests between all possible combinations of school geographical location groups are shown in Table 7.12. On peer schools' imitation of their development strategy, participants from District A reported lower values than participants from the three other districts, and participants from Districts B and D reported higher values than participants from District C. With regard to the consistency of implementation, participants from District A reported lower values than participants from the other three districts, and participants from District B reported higher values than participants from Districts C and D. Concerning future strategy, participants from District

Table 7.11 Participants' perceptions of strategic direction and intent by location

Item	Location	N	M	SD	F	P
Peer schools in the district wish to imitate our development strategy.	District A	352	3.28	1.06	31.36	<0.001
	District B	969	3.83	1.08		
	District C	130	3.51	1.11		
	District D	481	3.92	1.03		
All school work is conducted with the guidance of school development strategies.	District A	357	3.78	0.96	87.70	<0.001
	District B	974	4.57	0.69		
	District C	133	4.33	0.80		
	District D	484	4.33	0.84		
There is a clear strategy for the future.	District A	356	3.79	1.00	112.21	<0.001
	District B	973	4.66	0.63		
	District C	133	4.36	0.91		
	District D	485	4.47	0.76		
Our strategic direction is unclear to me.	District A	356	3.65	1.21	60.31	<0.001
	District B	1010	4.56	0.99		
	District C	132	4.22	1.10		
	District D	484	4.14	1.32		

Table 7.12 Participants' perceptions of strategic direction and intent by location: Post-hoc test

Item	District A vs. B	District A vs. C	District A vs. D	District B vs. C	District B vs. D	District C vs. D
Peer schools in the district wish to imitate our development strategy.	<0.001	0.034	<0.001	0.001	0.125	<0.001
All school work is conducted with the guidance of school development strategies.	<0.001	<0.001	<0.001	0.001	<0.001	0.976
There is a clear strategy for the future.	<0.001	<0.001	<0.001	<0.001	<0.001	0.152
Our strategic direction is unclear to me.	<0.001	<0.001	<0.001	0.001	<0.001	0.451

A reported the lowest values, while participants from District B reported higher values than those from Districts C and D. Regarding the clarity of strategic direction, participants from District A reported the lowest values, while participants from District B reported higher values than participants from Districts C and D.

7.2 Goals and Objectives

In order to understand the participants' perceptions of the goals and objectives of their school culture, four items—"Leaders set goals that are ambitious, but realistic;" "The leadership has 'gone on record' about the objectives we are trying to meet;" "Leaders continuously track our progress against our stated goals;" "I will devote myself to fulfilling the school's development goals"—were asked that required five-point Likert-type responses.

Overall findings

Table 7.13 shows a summary of the descriptive data of the participants' perception on these four items. In general, participants reported higher values than the mean on all four items, reporting especially high values on the last one. Participants largely felt positively about their school's goals and objectives.

Influence of Gender

Both male and female participants had favorable opinions of their school's goals and objectives (see Table 7.14). Males and females had similar opinions on the ambitiousness and feasibility of goals set by leaders, the public status of objectives, and leaders tracking progress toward stated goals. Females expressed greater commitment to fulfilling the school's development goals than males.

Influence of Educational Level

All four educational level groups held positive opinions of their school's goals and objectives (see Table 7.15). There were no significant differences among the four educational level groups on these items.

Table 7.13 Participants' perceptions of goals and objectives

Item	N	M	SD
Leaders set goals that are ambitious, but realistic.	1946	4.26	0.89
The leadership has "gone on record" about the objectives we are trying to meet.	1945	4.48	0.81
Leaders continuously track our progress against our stated goals.	1981	4.28	0.91
I will devote myself to fulfilling the school's development goals.	1945	4.51	0.74

Table 7.14 Participants' perceptions of goals and objectives by gender

Item	Gender	N	M	SD	t	P
Leaders set goals that are ambitious, but realistic.	Male	352	4.23	0.89	−0.72	0.472
	Female	1579	4.27	0.88		
The leadership has "gone on record" about the objectives we are trying to meet.	Male	350	4.43	0.82	−1.39	0.165
	Female	1580	4.50	0.80		
Leaders continuously track our progress against our stated goals.	Male	353	4.26	0.90	−0.55	0.581
	Female	1613	4.29	0.91		
I will devote myself to fulfilling the school's development goals.	Male	349	4.43	0.78	−2.38	0.018
	Female	1581	4.53	0.73		

Table 7.15 Participants' perceptions of goals and objectives by educational level

Item	Job title	N	M	SD	F	P
Leaders set goals that are ambitious, but realistic.	Postgraduate degree	74	4.27	0.78	0.19	0.902
	Bachelor's degree	1498	4.26	0.89		
	Diploma	276	4.28	0.85		
	Other	36	4.17	1.06		
The leadership has "gone on record" about the objectives we are trying to meet.	Postgraduate degree	74	4.31	0.78	1.46	0.225
	Bachelor's degree	1502	4.49	0.81		
	Diploma	271	4.52	0.75		
	Other	36	4.42	1.08		
Leaders continuously track our progress against our stated goals.	Postgraduate degree	74	4.28	0.79	0.53	0.660
	Bachelor's degree	1529	4.28	0.92		
	Diploma	282	4.30	0.87		
	Other	34	4.47	0.83		
I will devote myself to fulfilling the school's development goals.	Postgraduate degree	73	4.41	0.74	1.10	0.347
	Bachelor's degree	1500	4.50	0.74		
	Diploma	275	4.57	0.72		
	Other	35	4.51	0.78		

The P-values of post-hoc tests between all possible combinations of educational level groups are shown in Table 7.16.

Influence of Job Duty

All three job duty groups had positive perceptions of their school's goals and objectives (see Table 7.17). Participants of different job duty groups differed in their perceptions of the ambitiousness and feasibility of goals set by leaders, the public status of objectives, and their own commitment to fulfilling the school's development goals. However, there was no significant difference in how the three duty groups viewed the tracking of progress toward stated goals.

Table 7.16 Participants' perceptions of goals and objectives by educational level: Post-hoc test

Item	Postgraduate vs. Bachelor	Postgraduate vs. Diploma	Postgraduate vs. Other	Bachelor vs. Diploma	Bachelor vs. Other	Diploma vs. Other
Leaders set goals that are ambitious, but realistic.	0.935	0.915	0.563	0.717	0.523	0.458
The leadership has "gone on record" about the objectives we are trying to meet.	0.066	0.043	0.517	0.482	0.606	0.452
Leaders continuously track our progress against our stated goals.	0.947	0.906	0.322	0.719	0.219	0.296
I will devote myself to fulfilling the school's development goals.	0.297	0.100	0.496	0.163	0.931	0.669

Table 7.17 Participants' perceptions of goals and objectives by job duty

Item	Job duty	N	M	SD	F	P
Leaders set goals that are ambitious, but realistic.	Teacher	1677	4.26	0.89	4.76	0.009
	Middle leadership	125	4.50	0.68		
	Principal	38	4.18	0.77		
The leadership has "gone on record" about the objectives we are trying to meet.	Teacher	1678	4.48	0.82	5.12	0.006
	Middle leadership	124	4.72	0.55		
	Principal	38	4.55	0.55		
Leaders continuously track our progress against our stated goals.	Teacher	1711	4.28	0.92	1.49	0.226
	Middle leadership	125	4.42	0.86		
	Principal	38	4.24	0.68		
I will devote myself to fulfilling the school's development goals.	Teacher	1678	4.52	0.74	6.40	0.002
	Middle leadership	124	4.73	0.59		
	Principal	38	4.32	0.70		

The P-values of post-hoc tests between all possible combinations of job duty groups are shown in Table 7.18. Teachers and principals reported lower values than middle leadership on the ambitiousness and feasibility of goals

Table 7.18 Participants' perceptions of goals and objectives by job duty: Post-hoc test

Item	Teacher vs. Middle leadership	Teacher vs. Principal	Middle leadership vs. Principal
Leaders set goals that are ambitious, but realistic.	0.003	0.600	0.049
The leadership has "gone on record" about the objectives we are trying to meet.	0.002	0.592	0.263
Leaders continuously track our progress against our stated goals.	0.092	0.760	0.266
I will devote myself to fulfilling the school's development goals.	0.002	0.092	0.002

set by leaders. With regard to the publicity of objectives, middle leadership reported higher values than teachers. Concerning personal commitment to fulfilling the school's development goals, teachers and principals reported lower values than middle leadership.

Influence of Job Title

All three job title groups took a favorable view of their school's goals and objectives (see Table 7.19). Nonetheless, the groups showed statistically significant differences in their responses to all four items.

Table 7.19 Participants' perceptions of goals and objectives by job title

Item	Job title	N	M	SD	F	P
Leaders set goals that are ambitious, but realistic.	Junior	414	4.24	0.88	18.25	<0.001
	Middle	457	4.09	0.92		
	Senior	978	4.38	0.83		
The leadership has "gone on record" about the objectives we are trying to meet.	Junior	416	4.45	0.81	25.49	<0.001
	Middle	455	4.30	0.88		
	Senior	979	4.61	0.73		
Leaders continuously track our progress against our stated goals.	Junior	422	4.23	0.92	15.01	<0.001
	Middle	464	4.14	0.91		
	Senior	998	4.40	0.88		
I will devote myself to fulfilling the school's development goals.	Junior	416	4.51	0.72	18.14	<0.001
	Middle	454	4.35	0.79		
	Senior	978	4.60	0.70		

The P-values of post-hoc tests between all possible combinations of job title groups are shown in Table 7.20. On the ambitiousness and feasibility of goals set by leaders, the Senior group reported higher values than the Junior group, which in turn reported higher values than Middle group. With regard to the public status of objectives, the Senior group reported higher values than the Junior group, which in turn reported higher values than the Middle group. On leaders tracking progress toward stated goals, the Senior group reported the highest values. Regarding each participant's own commitment to fulfilling the school's development goals, the Senior group reported higher values than the Junior group, which in turn reported higher values than the Middle group.

Influence of Number of Years in Teaching

All three groups, those with less than 6 years of teaching experience, 6–15 years of experience, and more than 15 years of teaching experience, held positive perceptions of their school's goals and objectives (see Table 7.21). There were no significant differences among the three groups with different amounts of experience on the ambitiousness and feasibility of goals set by leaders or the personal commitment of participants to fulfilling the school's development goals. Groups with different amounts of experience differed in their perceptions of the public status of objectives and on tracking progress toward stated goals.

The P-values of post-hoc tests between all possible combinations of teaching year groups are shown in Table 7.22. Concerning the publicity of objectives and tracking progress, the group with more than 15 years of

Table 7.20 Participants' perceptions of goals and objectives by job title: Post-hoc test

Item	Junior vs. Middle	Junior vs. Senior	Middle vs. Senior
Leaders set goals that are ambitious, but realistic.	0.008	0.007	<0.001
The leadership has "gone on record" about the objectives we are trying to meet.	0.005	<0.001	<0.001
Leaders continuously track our progress against our stated goals.	0.150	0.001	<0.001
I will devote myself to fulfilling the school's development goals.	0.002	0.028	<0.001

Table 7.21 Participants' perceptions of goals and objectives by years in teaching

Item	Years in Teaching	N	M	SD	F	P
Leaders set goals that are ambitious, but realistic.	<6	374	4.28	0.82	0.33	0.717
	6–15	627	4.25	0.90		
	>15	910	4.28	0.89		
The leadership has "gone on record" about the objectives we are trying to meet.	<6	374	4.44	0.77	4.61	0.010
	6–15	626	4.44	0.84		
	>15	910	4.55	0.78		
Leaders continuously track our progress against our stated goals.	<6	381	4.22	0.92	4.46	0.012
	6–15	637	4.24	0.93		
	>15	927	4.35	0.88		
I will devote myself to fulfilling the school's development goals.	<6	375	4.49	0.72	1.49	0.225
	6–15	626	4.49	0.73		
	>15	909	4.55	0.75		

Table 7.22 Participants' perceptions of goals and objectives by years in teaching: Post-hoc test

Item	<6 vs. 6–15	<6 vs. >15	6–15 vs. >15
Leaders set goals that are ambitious, but realistic.	0.571	0.952	0.433
The leadership has "gone on record" about the objectives we are trying to meet.	0.971	0.022	0.008
Leaders continuously track our progress against our stated goals.	0.650	0.012	0.017
I will devote myself to fulfilling the school's development goals.	0.942	0.182	0.137

experience reported higher values than the group with less than 6 years of experience and the group with 6–15 years.

Influence of the School's Geographical Location

All four school geographical location groups had favorable opinions of their school's goals and objectives (see Table 7.23). However, there were significant differences among the four school geographical location groups on all four items being evaluated.

The P-values of post-hoc tests between all possible combinations of school geographical location groups are shown in Table 7.24. On the ambitiousness and feasibility of goals set by leaders, participants from

Table 7.23 Participants' perceptions of goals and objectives by location

Item	Location	N	M	SD	F	P
Leaders set goals that are ambitious, but realistic.	District A	355	3.69	0.99	73.74	<0.001
	District B	975	4.46	0.77		
	District C	132	4.18	0.91		
	District D	484	4.30	0.84		
The leadership has "gone on record" about the objectives we are trying to meet.	District A	353	3.89	0.96	105.88	<0.001
	District B	975	4.71	0.63		
	District C	133	4.43	0.82		
	District D	484	4.47	0.78		
Leaders continuously track our progress against our stated goals.	District A	355	3.75	0.97	58.79	<0.001
	District B	1009	4.46	0.83		
	District C	133	4.24	0.91		
	District D	484	4.31	0.87		
I will devote myself to fulfilling the school's development goals.	District A	353	4.01	0.88	85.67	<0.001
	District B	975	4.70	0.59		
	District C	132	4.46	0.75		
	District D	485	4.51	0.72		

District A reported lower values than participants from the other three districts. Participants from District B reported higher values than participants from Districts C and D. With regard to the public status of objectives, participants from District A reported the lowest values, while participants from District B reported higher values than participants from Districts C and D. Concerning leaders tracking goal progress, participants from District A reported the lowest values, while participants from District B reported higher values than participants from Districts C and D. Regarding personal commitment to fulfilling the school's development goals, participants from District A reported the lowest values, while participants from District B reported higher values than those from Districts C and D.

7.3 Vision

In order to understand participants' perceptions of vision in their school culture, four items—"Teachers have a shared vision of what the school will be like in the future;" "Leaders have a long-term viewpoint;" "Short-term thinking often compromises our long-term vision;" "Our vision

Table 7.24 Participants' perceptions of goals and objectives by location: Post-hoc test

Item	District A vs. B	District A vs. C	District A vs. D	District B vs. C	District B vs. D	District C vs. D
Leaders set goals that are ambitious, but realistic.	<0.001	<0.001	<0.001	<0.001	<0.001	0.168
The leadership has "gone on record" about the objectives we are trying to meet.	<0.001	<0.001	<0.001	<0.001	<0.001	0.601
Leaders continuously track our progress against our stated goals.	<0.001	<0.001	<0.001	0.006	0.001	0.446
I will devote myself to fulfilling the school's development goals.	<0.001	<0.001	<0.001	<0.001	<0.001	0.508

creates excitement and motivation for teachers"—were asked that required five-point Likert-type responses.

Overall findings

Table 7.25 shows a summary of the descriptive data of participants' responses to these items. In general, participants reported higher values than the mean on all four of these items. Participants generally held a positive opinion of their school's vision.

Influence of Gender

Both male and female participants held positive opinions of their school's vision (see Table 7.26). Females agreed more strongly with statements regarding teachers' shared vision of the school's future and compromising long-term vision with short-term thinking. Males and females had similar opinions on the long-term viewpoint of leaders and the excitement and motivation created by teachers' vision.

Influence of Educational Level

All four educational level groups took a favorable view of their school's vision (see Table 7.27). There were no significant differences among the four educational level groups on teachers' shared vision of the school's future, the long-term viewpoint of leaders, or the excitement and motivation created by

Table 7.25 Participants' perceptions of vision

Item	N	M	SD
Teachers have a shared vision of what the school will be like in the future.	1947	4.37	0.83
Leaders have a long-term viewpoint.	1983	4.45	0.83
Short-term thinking often compromises our long-term vision.	1942	4.21	1.15
Our vision creates excitement and motivation for teachers.	1978	4.16	1.00

Table 7.26 Participants' perceptions of vision by gender

Item	Gender	N	M	SD	T	P
Teachers have a shared vision of what the school will be like in the future.	Male	349	4.25	0.84	−3.04	0.002
	Female	1583	4.39	0.82		
Leaders have a long-term viewpoint.	Male	353	4.47	0.78	0.34	0.731
	Female	1615	4.45	0.83		
Short-term thinking often compromises our long-term vision.	Male	347	4.09	1.22	−2.14	0.032
	Female	1580	4.23	1.14		
Our vision creates excitement and motivation for teachers.	Male	352	4.13	1.03	−0.60	0.549
	Female	1611	4.17	0.99		

teachers' vision. Participants of different educational level groups differed in their perceptions of the prevalence of compromising long-term vision with short-term thinking.

The P-values of post-hoc tests between all possible combinations of educational level groups are shown in Table 7.28. Concerning the prevalence of compromising long-term vision with short-term thinking, participants with diplomas reported higher values than the other three educational level groups, and participants with bachelor's degrees reported higher values than those with postgraduate education.

Influence of Job Duty

All three job duty groups had positive perceptions of their school's vision (see Table 7.29). Teachers, middle leadership, and principals had similar opinions on teachers' shared vision of the school's future and the excitement and motivation created by teachers' vision. Participants of different job duty groups differed in their perceptions of the long-term viewpoint of

Table 7.27 Participants' perceptions of vision by educational level

Item	Job title	N	M	SD	F	P
Teachers have a shared vision of what the school will be like in the future.	Postgraduate degree	74	4.22	0.90	0.97	0.406
	Bachelor's degree	1500	4.37	0.82		
	Diploma	275	4.40	0.79		
	Other	36	4.33	0.96		
Leaders have a long-term viewpoint.	Postgraduate degree	74	4.41	0.74	1.21	0.305
	Bachelor's degree	1529	4.44	0.84		
	Diploma	282	4.54	0.74		
	Other	36	4.50	0.91		
Short-term thinking often compromises our long-term vision.	Postgraduate degree	73	3.70	1.48	8.21	<0.001
	Bachelor's degree	1497	4.21	1.15		
	Diploma	274	4.41	0.99		
	Other	36	3.94	1.29		
Our vision creates excitement and motivation for teachers.	Postgraduate degree	74	4.15	0.90	0.22	0.882
	Bachelor's degree	1524	4.15	1.01		
	Diploma	282	4.19	0.97		
	Other	36	4.22	0.90		

Table 7.28 Participants' perceptions of vision by educational level: Post-hoc test

Item	Postgraduate vs. Bachelor	Postgraduate vs. Diploma	Postgraduate vs. Other	Bachelor vs. Diploma	Bachelor vs. Other	Diploma vs. Other
Teachers have a shared vision of what the school will be like in the future.	0.117	0.095	0.484	0.626	0.792	0.666
Leaders have a long-term viewpoint.	0.727	0.225	0.571	0.071	0.662	0.807
Short-term thinking often compromises our long-term vision.	<0.001	<0.001	0.290	0.008	0.175	0.023
Our vision creates excitement and motivation for teachers.	0.984	0.742	0.716	0.484	0.651	0.862

Table 7.29 Participants' perceptions of vision by job duty

Item	Job duty	N	M	SD	F	P
Teachers have a shared vision of what the school will be like in the future.	Teacher	1680	4.38	0.82	0.55	0.578
	Middle leadership	124	4.44	0.71		
	Principal	37	4.30	0.70		
Leaders have a long-term viewpoint.	Teacher	1715	4.45	0.83	3.42	0.033
	Middle leadership	124	4.64	0.63		
	Principal	38	4.34	0.67		
Short-term thinking often compromises our long-term vision.	Teacher	1678	4.20	1.16	4.84	0.008
	Middle leadership	122	4.48	0.98		
	Principal	37	4.54	0.51		
Our vision creates excitement and motivation for teachers.	Teacher	1710	4.17	1.00	1.35	0.261
	Middle leadership	125	4.25	0.89		
	Principal	37	3.95	0.81		

leaders and the prevalence of compromising long-term vision with short-term thinking.

The P-values of post-hoc tests between all possible combinations of job duty groups are shown in Table 7.30. With regard to the long-term viewpoint of leaders, middle leadership reported higher values than teachers. Concerning the prevalence of compromising long-term vision with short-term thinking, teachers reported lower values than middle leadership.

Influence of Job Title

All three job title groups had favorable opinions of their school's vision (see Table 7.31). Nonetheless, the groups showed statistically significant differences in their responses to all four items.

The P-values of post-hoc tests between all possible combinations of job title groups are shown in Table 7.32. On teachers' shared vision of the school's future, the Senior group reported higher values than the Junior group, which in turn reported higher values than the Middle group. With regard to the long-term viewpoint of leaders, the Senior group reported higher values than the Junior group, which in turn reported higher values than the Middle group. Concerning the prevalence of compromising long-term vision with short-term thinking,

Table 7.30 Participants' perception on vision by job duty: Post-hoc test

Item	Teacher vs. Middle leadership	Teacher vs. Principal	Middle leadership vs. Principal
Teachers have a shared vision of what the school will be like in the future.	0.413	0.534	0.338
Leaders have a long-term viewpoint.	0.014	0.417	0.052
Short-term thinking often compromises our long-term vision.	0.009	0.071	0.761
Our vision creates excitement and motivation for teachers.	0.388	0.174	0.102

Table 7.31 Participants' perceptions of vision by job title

Item	Job title	N	M	SD	F	P
Teachers have a shared vision of what the school will be like in the future.	Junior	417	4.33	0.82	16.66	<0.001
	Middle	455	4.22	0.87		
	Senior	978	4.48	0.77		
Leaders have a long-term viewpoint.	Junior	422	4.45	0.79	21.47	<0.001
	Middle	465	4.26	0.89		
	Senior	999	4.56	0.77		
Short-term thinking often compromises our long-term vision.	Junior	416	4.19	1.18	17.40	<0.001
	Middle	452	3.97	1.26		
	Senior	977	4.35	1.06		
Our vision creates excitement and motivation for teachers.	Junior	423	4.17	1.00	14.25	<0.001
	Middle	463	3.97	1.05		
	Senior	996	4.26	0.94		

the Senior group reported higher values than the Junior group, which in turn reported higher values than the Middle group. Regarding the excitement and motivation created by teachers' vision, the Senior and Junior groups reported higher values than the Middle group.

Influence of Years in Teaching

All three teaching year groups felt positively about their school's vision (see Table 7.33). There were no significant differences among the three groups with different levels of experience on teachers' shared vision, the long-term

Table 7.32 Participants' perceptions of vision by job title: Post-hoc test

Item	Junior vs. Middle	Junior vs. Senior	Middle vs. Senior
Teachers have a shared vision of what the school will be like in the future.	0.035	0.003	<0.001
Leaders have a long-term viewpoint.	0.001	0.018	<0.001
Short-term thinking often compromises our long-term vision.	0.005	0.016	<0.001
Our vision creates excitement and motivation for teachers.	0.003	0.096	<0.001

Table 7.33 Participants' perceptions of vision by years in teaching

Item	Years in Teaching	N	M	SD	F	P
Teachers have a shared vision of what the school	<6	376	4.31	0.84	1.65	0.192
	6–15	627	4.40	0.76		
will be like in the future.	>15	909	4.38	0.85		
Leaders have a long-term viewpoint.	<6	379	4.46	0.76	0.89	0.413
	6–15	637	4.43	0.82		
	>15	931	4.48	0.84		
Short-term thinking often compromises our long-term vision.	<6	374	4.02	1.29	6.14	0.002
	6–15	627	4.22	1.16		
	>15	906	4.27	1.09		
Our vision creates excitement and motivation for teachers.	<6	380	4.19	0.96	0.23	0.797
	6–15	636	4.15	1.01		
	>15	926	4.16	1.00		

viewpoint of leaders, or the excitement and motivation created by teachers' vision. Groups with different levels of experience differed in their perceptions of the prevalence of compromising long-term vision with short-term thinking.

The P-values of post-hoc tests between all possible combinations of teaching year groups are shown in Table 7.34. With regard to compromising long-term vision with short-term thinking, the group with less than 6 years of experience reported lower values than both the group with 6–15 years of experience and the group with more than 15 years of experience.

Table 7.34 Participants' perceptions of vision by years in teaching: Post-hoc test

Item	<6 vs. 6–15	<6 vs. >15	6–15 vs. >15
Teachers have a shared vision of what the school will be like in the future.	0.082	0.120	0.728
Leaders have a long-term viewpoint.	0.493	0.696	0.184
Short-term thinking often compromises our long-term vision.	0.007	0.001	0.481
Our vision creates excitement and motivation for teachers.	0.502	0.624	0.790

Influence of the School's Geographical Location

All four school geographical location groups held positive perceptions of their school's vision (see Table 7.35). However, there were significant differences among the four groups on all four items being evaluated.

The P-values of post-hoc tests between all possible combinations of school geographical location groups are shown in Table 7.36. On the shared vision of teachers, participants from District A reported lower values than participants from the other districts, and participants from District B reported

Table 7.35 Participants' perceptions of vision by location

Item	Location	N	M	SD	F	P
Teachers have a shared vision of what the school will be like in the future.	District A	355	3.85	0.91	70.00	<0.001
	District B	976	4.55	0.71		
	District C	132	4.33	0.92		
	District D	484	4.38	0.79		
Leaders have a long-term viewpoint.	District A	356	3.86	1.01	95.80	<0.001
	District B	1009	4.66	0.65		
	District C	133	4.35	0.88		
	District D	485	4.47	0.77		
Short-term thinking often compromises our long-term vision.	District A	353	3.74	1.18	40.37	<0.001
	District B	974	4.46	0.99		
	District C	133	4.16	1.09		
	District D	482	4.05	1.31		
Our vision creates excitement and motivation for teachers.	District A	355	3.57	1.13	58.53	<0.001
	District B	1009	4.35	0.90		
	District C	131	4.04	1.00		
	District D	483	4.22	0.91		

Table 7.36 Participants' perceptions of vision by location: Post-hoc test

Item	District A vs. B	District A vs. C	District A vs. D	District B vs. C	District B vs. D	District C vs. D
Teachers have a shared vision of what the school will be like in the future.	<0.001	<0.001	<0.001	0.002	<0.001	0.447
Leaders have a long-term viewpoint.	<0.001	<0.001	<0.001	<0.001	<0.001	0.129
Short-term thinking often compromises our long-term vision.	<0.001	<0.001	<0.001	0.004	<0.001	0.333
Our vision creates excitement and motivation for teachers.	<0.001	<0.001	<0.001	0.001	0.013	0.060

higher values than participants from Districts C and D. With regard to the long-term viewpoint of leaders, participants from District A again reported lower values than participants from all other districts, and participants from District B reported higher values than participants from Districts C and D. Concerning the prevalence of compromising long-term vision with short-term thinking, participants from District A reported the lowest values, while participants from District B reported higher values than those from Districts C and D. Regarding the excitement and motivation created by teachers' vision, participants from District A reported the lowest values, while participants from District B reported higher values than those from Districts C and D.

7.4 Summary

The purpose of this chapter was to investigate school staff members' perceptions of mission, one of the school culture traits. There have been three sections in this chapter, one for each of the three indexes for mission: strategic direction and intent, goals and objectives, and vision. Four items were used for each area, using a five-point Likert-type item, which the participants then responded to. First section reported how participants responded to questions about the intention of peer schools to imitate their development strategy, the consistency of applying the development strategy to school work, and the clarity of the school's future strategy and strategic direction. Second section presented participants' perceptions of the ambitiousness and feasibility of goals set by leaders, how widely objectives were publicized, how well leaders tracked progress toward their stated goals, and the commitment of participants themselves to fulfilling the school's development goals. Third, participants responded items related to teachers' shared vision of the school's future, the long-term viewpoint of leaders, the prevalence of compromising long-term vision with short-term thinking, and the excitement and motivation created by teachers' vision.

The overall findings of the participants' responses to the items were recorded. The effects of variables such as gender, educational level, job duty, job title, years in teaching, and school location were considered for each item.

The findings presented in this chapter can be summarized as follows:

1. Participants in this investigation generally agree on the strategic direction and intent of their school culture. They believe peer schools wish to imitate their development strategy and that all their school work is conducted in accordance with school development strategies. They agree that there is a clear strategy for the future, and that the strategic direction is clear. However, significant differences can be observed between some groups in their perceptions of their school's strategic direction; for example, the Senior group reported a higher degree of agreement than the Junior and Middle groups. Participants from District A reported a lower degree of agreement than participants from the other three districts, and participants from District B reported a higher degree of agreement than District C.
2. Findings from the goals and objectives section showed a positive overall picture of staff members' agreement on this school culture index. Participants agreed that school leaders set ambitious and realistic goals. They thought that the leadership had "gone on record" about the objectives they were trying to meet. They thought leaders continuously tracked progress toward their stated goals. Participants expressed devotion to fulfilling the school's development goals. The factor of educational level did not have an effect on participants' perceptions of goals and objectives. However, differences can be found among job title groups and school location groups. Concerning the effect of job title, the Senior group reported a higher degree of agreement than the Junior and Middle groups. Participants from District A reported a lower degree of agreement than

participants from the other three districts, and participants from District B reported a higher degree of agreement than participants from Districts C and D. There was no difference between participants from Districts C and D.

3. Generally, participants agreed on the vision of their school culture. They thought teachers had a shared vision of what the school would be like in the future. They also thought the leaders had a long-term viewpoint. They agreed that short-term thinking did not compromise long-term vision and that teachers' vision created excitement and motivation for teachers. Participants' job title groups and school location groups were influential factors. The Junior and Senior groups reported higher levels of agreement than the Middle group. There were significant differences among school location groups. Specifically, participants from District A reported a lower degree of agreement than participants from the other three districts, and participants from District B reported a higher degree of agreement than participants from Districts C and D. There was no difference between participants from District C and District D.

In conclusion, the participants strongly agreed that their schools have good cultures concerning mission. The extent of agreement differed among groups based on job duty. Furthermore, school location and the staff members' job titles played an important role in their perceptions of their school's mission. The results also indicate further need to enhance involvement for participants from District A and participants in the Middle job title group.

8

Discussion, Reflection and Further Perspectives on School Culture Development in China

8.1 Summary of the Study

In earlier chapters of this book, we have introduced the background of this research project and, in particular, this pilot study. We have explained why it is important to give attention to school culture in the current education reform movement in China, presented a conceptual understanding of school culture and relevant studies on this topic, and explained why we focus on teachers and school principals' perspectives and how the research was conducted.

To generate the data for this pilot study, a survey-based investigation was conducted among 1,922 teachers and principals from 37 Chinese primary and middle schools (see chapter 3 for details of the survey). The Denison Organizational Culture Survey was used as an instrument for data generation, with minor revisions to fit Chinese school contexts. The survey used in this study is comprised of four traits with twelve indexes: involvement (empowerment, team orientation, capability development), consistency (core values, agreement, coordination and integration), adaptability (creating change, customer focus, organizational learning), and mission (strategic direction and intent, goals and objectives, vision). Four items were used to evaluate each index. The findings of the study have been reported in the body of this book. In this chapter, we will summarize the main findings and discuss the findings in relation to the research questions, literature, and Chinese school contexts. Limitations of this study and implications for future work will also be addressed. Finally, we will draw conclusions from this chapter and this book.

The major goal of this study has been to understand how teachers and principals perceive the current school culture at their schools and the current reform efforts with regards to school culture development. While conducting this pilot study, we assumed that a number of factors could potentially impact the teachers and principals' perceptions of the status of school culture in their

Table 8.1 Summary of findings of school culture index.

		Gender	Educational Level	Job Duty	Job Title	Years in Teaching	School Location
Involvement	Empowerment	2	0	2	4	0	4
	Team Orientation	2	0	3	4	1	4
	Capability Development	2	1	1	3	2	4
Consistency	Core Values	2	0	4	4	2	4
	Agreement	2	0	2	4	2	4
	Coordination & Integration	3	1	1	4	1	4
Adaptability	Creating Change	1	1	1	4	1	4
	Customer Focus	1	2	2	3	2	4
	Organizational Learning	2	1	3	4	1	4
Mission	Strategic Direction & Intent	2	1	3	4	2	4
	Goals & Objectives	1	4	3	4	2	4
	Vision	2	1	2	4	1	4

work environments. These factors include gender, educational level, job duty, job title, and years of teaching experience of the teachers and principals and the geographical location (by district) of the schools they work in. The main findings concerning each culture trait and index in relation to the impacts of gender, educational level, job duty, job title, years of teaching experience, and school location are summarized in Table 8.1. The numbers in the table indicate how many items (out of four) the above-mentioned factors significantly affect, for each school culture index.

In summary, in terms of their effects on the school culture indexes, school location by district within Beijing and job title of the respondent are the most influential factors. Gender and job duty have moderate effects. Educational level and years in teaching have the least effect on how teachers and principals think of the current school culture at their schools.

8.2 Major Findings and Discussion

Studies of educational change, and in particular of school culture development, very often report negative responses and resistance from staff to reform efforts (Fullan, 2001; Beaudoin and Taylor, 2004; Kruse and Louise, 2009; Preble and Gordon, 2011). However, the findings from our survey indicate that respondents generally have positive perceptions of the current state of school culture at their schools with regard to involvement, consistency, adaptation, and mission. The results indicate a general support for the ongoing reform efforts with regard to school culture establishment and improvement in Beijing Municipality. While we acknowledge that negative responses and resistance still exist, and the results of this study cannot be completely applied to other contexts in China, the positive results in this study (in particular, results of schools in Districts B, C, and D) indicate that it is likely that the school culture development campaign in Beijing has achieved a good level of success and satisfaction among both teachers and principals. This success can be used as a positive example for the expansion of this campaign on a larger scale in China.

The findings (as elaborated in Chapters 4–7) also show that factors such as school location, job title, job duty, educational level, and years in teaching make a difference (to varying degrees) in informants' perceptions of concrete aspects of their school cultures. Imbalances in the responses of different groups - for example, respondents from different districts, respondents with different job duties and titles, and female and male respondents - are observed, as discussed below.

Firstly, school location has the greatest impact on teachers and principals' perceptions of all items from the indexes of involvement (empowerment, team orientation, capability development), consistency (core values, agreement, coordination and integration), adaptability (creating change, customer focus, organizational learning), and mission (strategic direction and intent, goals and objectives, vision). The results indicate a significant geographical difference among districts in terms of how teachers and principals perceive the development of their school cultures. Schools in District A reported a considerably lower status for school culture development than schools in Districts B, C, and D. The differences can be attributed to the following interconnected reasons (as introduced in Chapter 3). The first is economic situation. District A is identified as an economically less developed area within Beijing Municipality, which means that is has fewer resources devoted to education in general and development projects for innovation in particular, compared with other areas.

As introduced in Chapter 3, District A had no experience in participating in school culture development projects at the time this investigation was conducted, whereas Districts B, C, and D all had a few ongoing projects related to school culture. The second is that, due to its comparatively less developed situation, school principals hired in this district are less qualified in terms of educational background and capabilities. The third reason is that District A is populated with migrant families, with 50 percent of the students' parents having low levels of education and holding low-paying temporary jobs such as those at construction sites and in service industries. The overall educational vision of this district is to provide basic education to all students, which means that it will take some time before the local government dives into further educational development projects such as improving school culture.

This finding echoes the conclusions of a previous study (Du et al., 2012), which shows how educational change is influenced by the institution's geographical location in China. Educational institutions located in less economically developed areas in China have fewer resources dedicated to education in general and development projects for innovation in particular, compared with other areas with better economic situation. This result also suggests that support from local government (in this study specifically the educational commission in each district) plays an important role in encouraging and enabling schools to initiate educational quality development projects like school culture improvement. Making educational innovation and change possible requires not only financial assistance, but also vision and strategy, with competent leadership to ensure the effective implementation of educational development actions.

Job title is the second most influential factor; it affects almost each item from every index. The Middle job title group of informants reported significantly lower values than both Senior and Junior groups on most items of the school culture indexes. On around half of the items, the Senior group reported higher values than Junior group, and the Junior group in turn reported higher values than the Middle group. In essence, the results show that staff in the Senior group feel most positively about the culture of their schools, the Junior group comes next, and the Middle job group respondents report the lowest values of the three groups. Interestingly, this finding is in line with a previous study on teaching staff's understanding of and attitude toward educational reform of teaching and learning in medicine and health higher education in China (Du et al., 2012). In that study, a 'U'-shaped tendency of attitudes was noted, in that Junior and Senior teaching staff were more supportive of educational reform than teaching staff with Middle job titles.

Thirdly, job duty has a significant effect on all the items under core values, and a partial effect on the other eleven indexes. Respondents were divided into three job duty groups in this study: teachers, middle leadership, and principals. Due to the specific focus of this study on understanding whether or not and in which ways teachers and principals have common perceptions of school culture, we mainly discuss (im) balances between teachers and principals.

Effective communication between teachers and principals has been identified by previous studies as one of the key factors that help to construct a great school culture (Beaudoin and Taylor, 2004; Elbot and Fulton, 2008; Kruse and Louise, 2009; Deal and Peterson,1999; 2009; Carter, 2011; Probe and Gordon, 2011). The present results showed that teachers and principals have similar perceptions of most items in the school culture indexes. This indicates that teachers and principals in this study demonstrate a basic common understanding of the status and quality of their school culture. This kind of common understanding demonstrates the existence of a certain level of positive communication between school leaders and teachers,

Furthermore, teachers and principals reported differently on several items. Teachers reported higher values than principals on teachers' beliefs regarding their impact on the school's development (empowerment), the consistency of their mode of dress and behavior with school culture (core values), the cooperative work required to achieve "win-win" solutions when disagreements occur (agreement), the difficulty of reaching agreement on key issues (agreement), the teachers' understanding of parents' wants and needs (customer focus), and of a clear future strategy that is well-communicated (strategic direction and intent). Principals reported lower values than teachers on taking into consideration the interests of parents in decision-making (customer focus), on the assurance of the school's future development (strategic direction and intent), and on the recent improvement of teachers' professional skills (capability development). The findings also show that principals give less attention to the impact of parents on school development practices (customer focus). This can be a serious issue and an obstacle to the establishment of a positive relationship between school and community (Beaudoin and Taylor, 2004; Kruse and Louise, 2009).

Although a certain level of common understanding and communication is demonstrated in the findings of this study, further improvement for more efficient communication is still needed. This is evident from the facts that-principals have a lower certainty of teachers' empowerment and involvement in school culture development than teachers themselves, and that they are less positive than teachers in assessing the levels of their common understanding of core values and the levels of agreement reached in cooperative work. Deal and Peterson (2009) stress the importance to leadership of recognizing the small successes of individuals and groups, which helps pave the way to bigger successes. The principals in this study reported a more negative assessment of teachers' understanding of the direction of school development and their progress in capability development. This can be seen as underestimating teachers' progress and achievements, which is not healthy for reaching the ultimate goal of improving school culture. These differences indicate a general lack of agreement on success criteria and norms of improvement, which is often regarded as obstacles to the success of educational development (Beaudoin and Taylor, 2004; Kruse and Louise, 2009; Deal and Peterson,1999; 2009; Carter, 2011; Probe and Gordon, 2011).

These results suggest that principals may need to communicate with teachers about these specific perspectives on their schools to achieve a common understanding of their school's culture. Participating in professional development activities/training programs for leadership may help principals better communicate with teachers about their vision and strategy, involve teachers in establishing the core values of the school, and collaboratively make correct decisions about which aspects of school culture to emphasize in future development (Deal and Peterson, 1999; 2009). This will help principals develop a functional professional community, facilitate organizational learning, and establish mutual trust within their schools (Kruse and Louise, 2009).

A previous study (Zembat, 2012) has shown that there is a positive relationship between preschool administrators' leadership styles and school culture in all sub-dimensions, namely, leadership based on cooperation, teacher cooperation, professional development, common goals, collegial support, and cooperative learning. This not only demonstrates the importance of the principal's role in school culture, but also indicates a possible way to improve school culture. Principals can change their school culture by changing their leadership style.

Fourthly, the results of this study also demonstrate that gender makes a significant impact on the results, with female respondents reporting higher values than males on the majority of index items. This pilot study has not succeeded in pinpointing relevant studies on school culture from a gender perspective. The above-mentioned study (Du et al., 2012) on teaching staff's understanding of and attitudes toward education reform of teaching and learning in medicine and health higher education in China reported no gender impact. The majority of school teachers in China are female, and there is a notable imbalance of gender participation in this study (with females representing 81.4% and males 17.8% of the total 1.992 respondents). A female-dominated teachers' culture can possibly be assumed although without scientific evidence yet. Nevertheless, this background cannot be used to explain the gender impact identified in this study.

Informants' educational backgrounds and years of teaching experience did not show a significant impact on the results of this pilot study.

8.3 Reflection on the Concepts and Instrument

In Chapter 2 of this book, our theoretical understanding of the concept of culture focuses on the complex nature of culture. This perspective presents the opportunity to emphasize the change and development aspects of culture. It allows us not only to see what culture is, but also to better understand how culture operates in relation to educational practice. In relation to the study of school culture, we hold the view that school culture is highly complex and is subject to ongoing negotiation and interaction among many actors such as principals, teachers, students, parents, and other possible actors. With regard to this pilot study, we believe that it is essential that principals and teachers share a common understanding and view of what a good school culture should be and how they can build a good school culture collaboratively.

In line with the above-mentioned conceptual understanding and view, this pilot study took an organizational perspective to studying and diagnosing school culture because school culture is often considered a branch of organizational culture in an educational context (Schoen and Teddlie, 2008).

The Denison Organizational Culture Survey is employed as an instrument in this study to investigate how teachers and principals perceive school culture development at their schools for two reasons. Firstly, the survey and its theoretical background are in line with the process approach to the conceptual understanding in this study, which places emphasis on the complex nature of culture and school culture. Secondly, the survey offers opportunities to understand and diagnose school culture and allows for the possibility of recommendations for further improvement. This aspect of the survey can be seen as an advantage that fits the research purpose of the present study. The choice of method is, at the same time, challenged by the fact that the Denison Organizational Culture Survey has not been employed in an educational context at the school level, nor, in particular, in a Chinese educational context.

In the field of education, the Denison Organizational Culture Survey has not often been used in the study of school culture, but rather has mainly been used in universities (Denison Consulting, 2010). To our knowledge, the present study represents the first time the Denison Organizational Culture Survey has been employed in the Chinese primary and middle school context. Results of this study provide important and thorough information for understanding organizational culture, specifically school culture in the Chinese context, and it enriches the theory related to organizational culture and school culture. The pilot study also proves the Denison Organizational Culture Survey can also be used in a Chinese school context, despite the fact that the instrument was developed in Western companies. This expands the Denison Organizational Culture Survey usage.

8.4 Recommendations

Based on the results of this investigation and reflective evaluation of this pilot study, a list of recommendations for the improvement and development of the school culture campaign within Beijing Municipality (which is expected to be expanded to a national scale in the long run) is summarized below. These recommendations are aimed at educational institutions to make decisions, develop strategies and take actions as well as sustain the educational innovations.

1. In order to further develop school culture in the educational change process in a balanced way and to secure educational equality in the long run, it is important that educational policy-making institutions provide sufficient and equally distributed support and resources in order to assist all schools in realizing their visions and goals. Equity in resource division and sufficient support from local government will enable schools from less economically developed areas to improve their school culture and achievement level. The necessary support includes not only financial assistance, but also vision and strategy, with competent leadership, to ensure the effective implementation of educational development and innovation actions.
2. From a sustainable development point of view, it is necessary for school leaders to develop internal strategies to achieve effective communication and common understanding among staff members from diverse groups. Certain groups of teachers, such as the Middle job title group, should be paid special attention via individual and/or small-scale conversations

and discussions with school leaders. This can be done through various ongoing activities such as seminars, workshops, social activities, and other professional communication development activities.

3. Positive relationships and effective communication between principals and teachers are essential for any educational development or cultural change. School principals should consider attending professional leadership development activities before implementing change actions. Well-communicated agreement between principals and teachers can not only help to identify the roles and responsibilities of each party in the school culture development process, but also clarify mutual expectations and evaluations with a common understanding of success criteria. External experts can be invited to assist teachers and principals in reaching a common understanding concerning current levels of individual capabilities and skills, as well as how to further improve them.
4. The current stage of school culture development has placed a major focus on the establishment of internal mechanisms. Further efforts should give weight to connections to the outside environment through building strong community relations and making better use of external resources.
5. Principals take a leading role in the school culture construction and improvement process. However, it is important that principals be mindful of the fact that school culture development is a crucial part of organizational learning in the knowledge society. This means that school improvement not only involves the individual development of leadership for principals, but also demands a holistic and comprehensive approach to school development that includes systematic institutional support and strategies, collective learning, and teamwork.
6. Students play a key role in any educational change, and in particular in the school cultural construction process. Sufficient attention should be paid to the involvement and influence of students in the process of developing a positive school culture with the ultimate goal of improving students' learning and producing responsible, productive and contributing citizens. This can be done by inviting students to participate in designing and organizing activities aimed at establishing symbolic values and school spirit and in evaluation activities aimed at assuring the quality of school culture development.
7. Establishing a research-based approach to educational change and reform is beneficial to its quality assurance and sustainability. Research that

includes multiple investigation methods can make long-lasting contributions to school culture development by providing scientific documentation, reflection, and evaluation as well as generating evidence-based recommendations for better practices.

8.5 Conclusions, Limitations, and Future Perspectives

Taking an organizational perspective to studying and diagnosing school culture, this study investigated teachers and principals' understanding and perceptions of school culture development in China. A survey was conducted among 1,992 informants, including 1,719 teachers and 38 principals (including vice principals), from 37 schools (located in four different districts) in Beijing Municipality. Findings were analyzed and discussed in relation to relevant literature on school culture development as well as the Chinese context. The results of this study indicated that the respondents had overall positive perceptions on school culture construction and development in the educational change process. In general, they showed a positive attitude toward the school culture improvement initiatives, reported satisfaction about their current school culture and held confidence in the direction their school culture is heading. A general common understanding and shared values between teachers and principals have been observed concerning their perceptions of the cultures of the schools they work in. However, this pilot study demonstrated that certain factors, such as school location and respondents' job duty, job title, and gender, make a difference in how participants perceive the school's culture. Participants from a less economically developed district had less awareness of, involvement in, and understanding of school culture development in all its aspects. They also reported less involvement in the decision-making and establishment of school culture process. Females reported higher values than males and Senior and Junior job title groups reported higher values than the Middle group on all aspects of the school culture survey. A notable gap between teachers and principals was identified in their perception of the core values and norms of the school as well as of communication between teachers and principals.

The empirical evidence of this study was primarily based on a study conducted with teachers and principals from 37 schools in Beijing. A quantitative method was used with the aim of gaining an overview of how teachers and principals understand and perceive their current school culture development and diagnosing the general status of this field in China. This method, however, limited the opportunity for in-depth understanding of informants' expectations, viewpoints, and experiences of the change process, which are the focuses of most studies in this field (Maslowski, 2006) and are also important aspects of the study of school culture development and educational change. Furthermore, due to time limitations, this pilot study did not include the perspectives of students, who play an essential role in educational change in general and school culture development in particular.

The results and limitations of the current pilot study also suggest a series of considerations for further research topics; for example, qualitative research on the life experiences of teachers, principals, students, and parents during the process of school culture change and development, including challenges, obstacles, and opportunities for better practices from both institutional and individual perspectives could be carried out. Professional development activities and evaluation of effects are also aspects that demand further attention. In addition, cross-geographical location (regions, countries) studies would provide beneficial contributions. Studies on how school culture development can improve school effectiveness would also add value to this field of research. As suggested by previous studies (Deal and Peterson, 2009; Guen, 2013), probing into school culture is the first step for the enhancement of the effectiveness of any school.

To conclude, this study suggests that to further enhance school culture development in China, it is important that teachers and principals share a common understanding of the school vision and mission and reach agreement on how to make changes in the daily practice of school culture in order to realize the goals. For that purpose, effective communication is of the essence. Educational change is to a large degree dependent on what teachers think and do (Fullan, 2007). Organizational effectiveness and improvement are independent on the staff's well-being and commitment (Deal and Peterson, 2009). Since individual teachers play an important role in the change process of education, it is essential to motivate teachers to become engaged in development and change activities (Fullan, 2007) and provide them with needed knowledge and skills. For this, it is important to offer institutionalized professional staff development. Finally, for educational development to be further carried forward successfully, there is a general need for better distribution of educational resources (both financial and leadership), common understanding, agreed-upon goals, and efficient communication between principals and teachers.

Contributors

Kai Yu (余凯), Ph.D., is an associate professor of education and public policy in the Faculty of Education at Beijing Normal University. He earned his Ph.D. in comparative education from Beijing Normal University in 1999 and an Ed. D. degree in education policy and leadership from the University of Massachusetts at Amherst in 2008. He worked as policy analyst for the Chinese Ministry of Education from 1999 to 2003 and for the United Nations Department of Economic and Social Affairs in 2006. At present, he serves as a consultant to UNICEF and Chinese governmental agencies and is a board member of the Zigen Rural Education and Development Fund of China. Among other works, Dr. Kai Yu is the author of *The Implementation of Inclusive Education in Beijing: Exorcizing the Haunting Specter of Meritocracy* and *American Education*, as well as numerous articles, book chapters, and monographs on education and public policy, school improvement, and evaluation studies.

Xiaoju Duan (段小菊), Ph.D., is a post-doctoral fellow in the Department of Learning and Philosophy at Aalborg University. She is also an assistant professor at the Institute of Psychology, Chinese Academy of Sciences. Her Ph.D. is in educational and developmental psychology. She has published a number of peer-reviewed articles in the fields of education, neurosciences, and cognitive development.

Xiangyun Du (杜翔云), Ph.D., is a professor in the Department of Learning and Philosophy and director of the Confucius Institute for Innovation and Learning at Aalborg University. She is also an adjunct professor at Beijing Normal University and at China Medical University. Her main research interests include innovative teaching and learning in education, particularly, problem-based and project-based learning method in fields ranging from engineering, medicine and health, and language education, to diverse social, cultural and educational contexts. She has also engaged with educational institutions in over 10 countries in substantial work on pedagogy development in teaching and learning. Professor Du has over 120 relevant international publications including monographs, international journal papers, edited books and book chapters, as well as conference contributions.

References

1. American heritage dictionary of the English language (3rd ed.). (1992). Boston: Houghton Mifflin.
2. Barnett, G. A. (1988). Communication and organizational culture. In G. M. Goldhaberand G. A. Barnett (eds), Handbook of organizational communication. New Jersey: Ablex. pp. 101–130.
3. Barth, R. (2002). The culture builder. Educational Leadership, 59(8), 6–11.
4. Beaudoin, M. N. and Taylor, M. (2004). Creating a positive school culture – how principals and teachers can solve problems together. Thousand Oaks: Corwin Press and SAGE, a joint publication.
5. Bu, Y. H. and Li, J. C. (2013). The new basic education and whole school reform: A Chinese experience. Frontier of Education in China, 8(4), 576 595.
6. Business Daily Update, May 8, 2013.
7. Cameron, K. S., and Quinn, R. E. (1999). Diagnosing and changing organizational culture. Reading MA: Addison-Wesley.
8. Carducci, L. (2012). Work in progress: Chinese education from a foreign expert's perspective. China Intercontinental Press, 27.
9. Carter, S. C. (2011). On purpose – how great school cultures from strong character. Thousand Oaks: Corwin Press and SAGE, a joint publication.
10. Chatman, J. (1991). Matching people and organizations: Selection and socialization in public accounting firms. Administrative Science Quarterly, 36(3), 459–485.
11. Cooke, R. A. and Rousseau, D. M. (1988). Behavioral norms and expectations: A quantitative approach to the assessment of organizational culture. Group and Organization Management, 13(3), 245–273.
12. Cui, Y. H. and Zhou, W. Y. (2007). Xuexiao wenhua jianshe: yizhong zhuanye de shijiao (School culture construction: a professional perspective). Jiaoyu fazhan yanjiu (Educational development and research), 5, 29–33. (in Chinese)

13. Deal, T. and Peterson, K. (1990). The principal's role in shaping school culture. Washington, D. C.: U. S. Department of Education.
14. Deal, T. and Peterson, K. (1999). Shaping school culture – pitfalls, paradoxes, and promises (2nd eds). San Francisco, CA: Jossey-Bass.
15. Deal, T. and Peterson, K. (2009). Shaping school culture – the heart of leadership. San Francisco, CA: Jossey-Bass.
16. Deng, Y. Z. and Xi, C. H. (2007). Xinkegai Beijingxia xuexiao wenhua yanjiu zongshu (Sum-up of school culture research in the new-round curriculum reform). Xiandai jiaoyu luncong (The Modern Education Journal),130(4), 32–37. (in Chinese)
17. Denison Consulting. (2010). University of North Texas Health Science Center: Culture's role in their vision to become a top ten health science center (Retrieved from http://www.denisonconsulting.com/resource-library/university-north-texas-health-science-center-cultures-role-their -vision-become-top).
18. Denison, D. R., Janovics, J., Young, J., and Cho, H. J. (2006). Diagnosing organizational cultures: Validating a model and method. Document of Denison Consulting Group.
19. Denison, D., Nieminen, L., and Kotrba, L. (2013). Diagnosing organizational cultures: A conceptual and empirical review of culture effectiveness surveys. European Journal of Work and Organizational Psychology, (ahead of print), 1–17.
20. Du, X. Y., Shi, J. N., Zhao, Y. H., and Sun, B. Z. (2012). Change and Reform in Medicine and Health Education in China: A Teaching Staff's Perspective. Aalborg: RiverPublisher.
21. Elbot, C. F. and Fulton, D. (2008). Building an intentional school culture. Thousand Oaks, CA: Corwin Press and SAGE.
22. Ember, C. R. and Ember, M. (1981). Cultural anthropology (3rd ed). Englewood Cliffs, NJ: Prentice-Hall.
23. Fey, C. F. and Denison, D. R. (2003). Organizational culture and effectiveness: Can American theory be applied in Russia? Organization Science, 14(6), 686–706.
24. Fullan, M. (1998). Leadership for the 21st century: breaking the bonds of dependency. Educational Leadership, 55(7), 6–10.
25. Fullan, M. (2001). Leading in a culture of change. San Francisco, CA: Jossey-Bass.
26. Fullan, M. (2005). Leadership and sustainability: system thinkers in action. Thousand Oaks, CA: Corwin Press.

27. Fullan, M. (2007). The new meaning of educational change. New York: Teachers College Press.
28. Garcia, S. B. and Guerra, P. L. (2006). Conceptualizing culture in education: implications for schooling in a culturally diverse society. In: Redefining culture – perspectives across the disciplines. Baldwin, J. R., Faulkner, S. L., Hecht, M. L. and Lindsley, S. L. (eds). New Jersey: Lawrence Erlbaum Associates, Publishers. 2006. pp. 103–115.
29. Gillespie, M. A., Denison, D. R., Haaland, S., Smerek, R., and Neale, W. S. (2008). Linking organizational culture and customer satisfaction: Results from two companies in different industries. European Journal of Work and Organizational Psychology, 17(1), 112–132.
30. Goffee, R., and Jones, G. (1998). The character of the corporation: How your company's culture can make or break your business. New York: Harper Business.
31. Gu, M. Y. (2006). Lun xuexiao wenhua jianshe (On the school culture construction). Journal of Southwest China Normal University (Humanities and Social Sciences Edition), 32(5), 67–70. (in Chinese)
32. Gudlaugsson, T. (2013). Psychometric properties of the Icelandic version of the Denison Organizational Culture Survey. International Journal of Business and Social Science, 4(4), 13–23.
33. Guen, B., and Caglayan, E. (2013). Implications from the diagnosis of a school culture at a higher education institution. Turkish Online Journal of Qualitative Inquiry, 4(1), 47–59.
34. Halpin, A. W. (1966). Theory and research in administration. New York: Macmillan.
35. Hecht, M. L., Baldwin, J. R., and Faulkner, S. L. (2005). The (in) conclusion of the matter: Shifting signs of models of culture. In J. R. Baldwin, S. L. Faulkner, M. L. Hecht and S. L. Lindsley (Eds.) Redefining culture: Perspective across the disciplines. USA: Lawrence Erlbaum Associates.
36. Heller, M. F. and Firestone, W. A. (1995). Who's in charge here? Sources of leadership for change in eight schools. The Elementary School Journal, 96(1), 65–86.
37. Hofstede, G., Bond, M. H., and Luk, C. (1993). Individual perceptions of organizational cultures: A methodological treatise on levels of analysis. Organization Studies, 14(4), 483–503.
38. Hoy, W. K., and Clover, S. I. R. (1986). Elementary school climate: A revision of the OCDQ. Educational Administration Quarterly, 22, 93–110.

39. Hoy, W. K., and Feldman, J. A. (1987). Organizational health: The concept and its measure. Journal of Research and Development in Education, 20, 30–38.
40. Hoy, W. K., Tarter, C. J., and Kottkamp, R. B. (1991). Open schools/healthy schools: measuring organizational climate. Beverly Hills, CA: Corwin Press.
41. Hu, Y. M., Zhang, Z., and Liang, W. Y. (2009). Efficiency of primary schools in Beijing, China: An evaluation by data envelopment analysis. International Journal of Educational Management, 23(1), 34–50.
42. Jensen, I. (2007). Introduction to cultural understanding. Roskilde: Roskilde University Press.
43. Jung, T., Scott, T., Davies, H. T. O., Bower, P., Whalley, D., McNally, R., et al. (2009). Instruments for exploring organizational culture: A review of the literature. Public Administration Review, 69(6), 1087–1096.
44. Kahn, J. S. (1989). Culture, demise or resurrection? Critique of Anthropology, 9(2), 5–25.
45. Keesing, R. M. (1981) Theories of culture. In: R. W. Casson (ed), Language, culture, and cognition. New York: Macmillan. pp. 42–67.
46. Kottkamp, R. B., Mulhern, J. A., and Hoy, W. K. (1987). Secondary school climate: A revision of the OCDQ. Educational Administration Quarterly, 23, 31–48.
47. Kroeber, A. L. and Kluckhohn, C. (1952). Culture: a critical view of concepts and definitions. Cambridge: Harvard University Press.
48. Kruse, S. D. and Louis, K. S. (2009). Building strong school culture – a guide to leading change. Thousand Oaks: Corwin Press and SAGE, a joint publication.
49. Li, L. Q. (2004). Education for 1.3 Billion. Foreign Language Teaching and Research Press.
50. Luo, X. M. (2011). Curriculum reform in the course of social transformation: The case of Shanghai. Cultural Studies, 25(1), 42–54.
51. Martin, J. (2002). Organizational culture: mapping the terrain. Thousand Oaks: SAGE.
52. Maslowski, R. (2001). School culture and school performance: an explorative study into the organizational culture of secondary schools and their effects. Twente: Twente University Press.
53. Maslowski, R. (2006). A review of inventories for diagnosing school culture. Journal of Educational Administration, 44(1), 6–35.
54. Muhammad, A. (2009). Transforming school culture – how to overcome staff division. Bloomington: Solution Three Press.

55. Preble, B. and Gordon, R. (2011). Transforming school climate and learning. Thousand Oaks: Corwin Press.
56. Rosaldo, R. I. (2006). Foreword: defining culture. In: Baldwin, J. R., Faulkner, S. L., Hecht, M. L. and Lindsley, S. L. (eds). Redefining culture – perspectives across the disciplines. New Jersey: Lawrence Erlbaum Associates Publishers. pp. 9–13.
57. Ryan, J. (2011). Education reform in China: Changing concepts, contexts and practices. London: Routledge.
58. Schein, E. H. (1985). Organizational culture and leadership (1st Edition). San Francisco: Jossey-Bass.
59. Schein, E. H. (1992). Organizational culture and leadership (2nd Edition). San Francisco: Jossey-Bass.
60. Schoen, L. T., and Teddlie, C. (2008). A new model of school culture: a response to a call for conceptual clarity. School Effectiveness and School Improvement, 19(2), 129–153.
61. Shi, Z. Y. (2005). Xuexiao wenhua de hexin: jiazhiguan jianshe (Core of school culture: construction of values). Jiaoyu kexue yanjiu (Educational science research), 8,17–22. (in Chinese)
62. Smircich, L. (1983). Concepts of culture and organizational analysis. Administrative Science Quarterly, 28, 339–358.
63. Steele, R. and Suozzo, A. (1994). Teaching French culture: Theory and practice. Lincolnwood, National Textbook Co.
64. Tan, G. Y. (2012). The One-Child policy ad privatization of education in China. International Education, Fall.
65. Tsang, M. C. (1996). Financial reform of basic education in China. Economics of Education Review, 15(4), 423–444.
66. Treiman, D. J. (2013). Trends in educational attainment in China. Chinese Sociological Review, 45(3), 3–25.
67. Van der Westhuizen, P. C., Oosthuizen, I., and Wolhuter, C. C. (2008). The relationship between an effective organizational culture and student discipline in a boarding school. Education and Urban Society, 40(2), 205–224.
68. Van Houtte, M., and Van Maele, D. (2011). The black box revelation: in search of conceptual clarity regarding climate and culture in school effectiveness research. Oxford Review of Education, 37(4), 505–524.
69. Wang, D. H. (2012). Shilun xinxingshixia xuexiao wenhua jianshe (On the cultural construction of primary and secondary schools in the new situation). Jiaoyu yanjiu (Educational research),384(1), 4–8. (in Chinese)

70. Wang, H. X. (2011). Access to higher education in China: Differences in opportunity. Frontier of Education in China, 6 (2), 227–247.
71. Wang, J. Y. and Zhao, Z. C. (2011). Basic education curriculum reform in rural China: Achievements, problems, and solutions. Chinese Education and Society, 44(6), 36–46.
72. Wu, X. L. (2013). Woguo jinsanshinian xuexiao wenhua yanjiu tanxi: guochengzhexue de shijiao (Exploration on school culture research in recent 30 years in China: the perspective of process philosophy). Jiaoyu fazhan yanjiu (Educational development and research),10, 16–22. (in Chinese)
73. Xin, T. and Kang, C. H. (2012). Qualitative advances of China's basic education since reform and opening up. Chinese Education and Society, 45(1), 42–50.
74. Yang, Q. (2009). Xuexiao wenhua jianshezhong de xiangguan yinsu fenxi (Analysis on the related factors in the construction of school culture). Jiaoyu yanjiu (Educational research), 348(1), 106–110. (in Chinese)
75. Yang, X. W. (2012). Changes in practice and reconstruction of theory. Chinese Education and Society, 45(1), 51–58.
76. Zembat, R., Sezer, A. A. O. T., and Biber, B. O. B. K. (2012). The relationship between preschool administrators leadership styles and school culture. e-Journal of New World Sciences Academy, 7(2), 798–811.
77. Zhang, J. T., Fang, Y. Y., and Ma, X., L. (2010). The latest progress report on ICT application in Chinese basic education. British Journal of Educational Technology, 41(4), 567–573.
78. Zhao, L. T. (2009). Between local community and central state: Financing basic education in China. International Journal of Educational Development, 29, 366–373.
79. Zhu, C. Devos, G. and Li, Y. F. (2011). Teacher perceptions of school culture and their organizational commitment and well-being in a Chinese school. Asia Pacific Educational Review, 12, 319–28.

Index

Appendix 1

Chinese Educational System

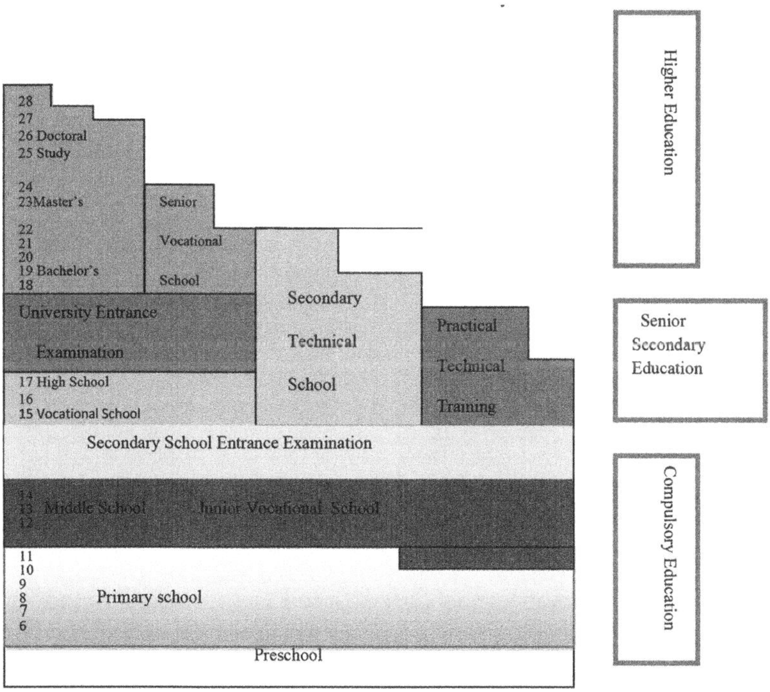

Figure is made based on information from open source, retrieved Jan 12th, 2014, http://www.harvest.org.hk/chi/articles/art020030.php

Appendix 2

Education and Human Resources Index in Beijing (2009–2020)

Indicators	2009	2015	2020
Gross Preschool Enrollment (%)	90.3	95	99
Gross Compulsory Education Enrollment (%)	105	Over 100	Over 100
Gross High School Enrollment (%)	98	99	99
Number of College Students per 100,000 residents	6369	6700	6700
Percentage of K-12 Schools with access to the Internet (%)	90.8	97	100
International Students (10,000)	7.1	12	18
Average Education Level of Incoming Labor Force (year)	14	15	15.5
Percentage of Age-group Labor Force with Higher Education (%)	35	40	48

http://www.bjedu.gov.cn/publish/portal0/tab103/info6764.htm. Retrieved Jan 12th, 2013

Appendix 3

Number of Schools in Beijing by Types and Levels

	Number of Schools
Total	**3314**
1.Higher Education	**179**
(Postgraduate Institutions)	(135)
(Universities and Colleges)	(56)
(Research Institutions)	(79)
General Universities	91
Supervised by Central Government	37
Supervised by Municipal Government	54
(Public)	39
(Private)	15
Adult Higher Education Institutions	19
Private Adult Higher Education Institutions	69
2.Secondary Education	**760**
High Schools	419
General High Schools	289
Secondary Vocational Schools	130
General Secondary Professional Schools	31
Adult Secondary Professional Schools	11
Vocational High Schools	54
Engineering Professional Schools	34
Middle Schools	341
3.Primary Education	**1081**
4.Reformatory School for Juvenile Delinquents	**6**
5.Special Education	**22**
6.Preschool Education	**1266**

Source: Beijing Commission of Education 2013

Appendix 4

Number of Primary Students by Districts and Grades in Beijing

	Number of Graduates	Number of Students newly enrolled	Number of Students in Schools							
			Total	Female	Grade 1	Grade 2	Grade 3	Grade 4	Grade 5	Grade 6
Total	**109492**	**141738**	**718655**	**331036**	**141999**	**135170**	**115765**	**105031**	**112607**	**108083**
Core Downtown										
Dongcheng	7962	8162	46721	21956	8162	8666	7559	7102	7825	7407
Xicheng	8680	11083	55636	25870	11084	10978	8671	7969	8626	8308
Expanded Core Districts										
Chaoyang	13181	21444	98530	44901	21456	19697	15831	13837	14160	13549
Fengtai	10277	12186	65935	29463	12188	11418	10961	10293	10556	10519
Shijingshan	3516	4240	21386	9920	4243	3928	3477	3153	3533	3052
Haidian	20942	25110	131547	60630	25114	25215	20759	19146	21005	20308
Developing Districts										
Fangshan	6800	8378	41979	19834	8407	7937	6765	6093	6791	5986
Tongzhou	7131	11889	54212	24548	11918	10132	8834	7825	7970	7533
Shunyi	5224	6890	35858	16464	6914	6865	6073	5201	5481	5324
Changping	5616	10356	50670	22954	10375	9202	8320	7537	7850	7386
Daxing	6404	8294	42418	19461	8368	7919	6876	6262	6621	6372
Ecological Districts										
Mentougou	2033	1946	11106	5262	1949	1942	1735	1666	1924	1890
Huairou	2555	3015	15483	7370	3021	2815	2525	2212	2554	2356
Pinggu	2925	2974	15542	7328	2983	2825	2431	2241	2471	2591
Miyun	3640	3739	19705	9377	3749	3679	3184	2802	3110	3181
Yanqing	2606	2032	11927	5698	2068	1952	1764	1692	2130	2321

Source: Beijing Commission of Education 2013
http://zfxxgk.beijing.gov.cn/fgdyna.prinfodetail.prPlanDetailInfo.do?GM_T_CATALOG_INFO/CATA_INFO_ID=374955 Retrieved Jan 12th, 2014

	Number of Schools	Number of Classes						
		Total	Grade 1	Grade 2	Grade 3	Grade 4	Grade 5	Grade 6
Total	**1081**	**21989**	**4124**	**3955**	**3583**	**3353**	**3524**	**3450**
Core Downtown								
Dongcheng	63	1433	253	248	242	227	235	228
Xicheng	72	1730	328	332	278	257	269	266
Expanded Core Districts								
Chaoyang	128	3358	672	620	528	495	523	520
Fengtai	87	1958	343	324	325	315	324	327
Shijingshan	31	670	126	120	112	104	109	99
Haidian	107	3693	681	677	597	562	604	572
Developing Districts								
Fangshan	108	1371	256	254	225	207	223	206
Tongzhou	81	1485	319	268	243	216	221	218
Shunyi	39	1014	186	188	174	149	157	160
Changping	99	1540	295	269	252	237	243	244
Daxing	88	1305	240	241	216	203	205	200
Ecological Districts								
Mentougou	35	386	64	65	62	60	67	68
Huairou	25	458	84	80	76	74	74	70
Pinggu	44	525	96	92	85	81	87	84
Miyun	40	594	104	103	96	92	100	99
Yanqing	34	469	77	74	72	74	83	89

Source: Beijing Commission of Education

Appendix 5

Number of Primary Teachers and Staff by Districts in Beijing

	Teachers and Staff	
	Total	Curriculum Teachers
Total	**55710**	**46783**
Core Downtown		
Dongcheng	4280	3455
Xicheng	4514	3823
Expanded Core Districts		
Chaoyang	7241	6553
Fengtai	4583	3902
Shijingshan	1600	1386
Haidian	7199	6644
Developing Districts		
Fangshan	3644	2925
Tongzhou	3584	3081
Shunyi	2763	2262
Changping	3388	2844
Daxing	3529	3012
Ecological Districts		
Mentougou	1506	1203
Huairou	1775	1195
Pinggu	2501	1735
Miyun	2092	1510
Yanqing	1511	1253

Source: Beijing Commission of Education 2013, http://zfxxgk.beijing.gov.cn/fgdyna.prinfodetail.prPlanDetailInfo.do?GM_T_CATALOG_INFO/CATA_INFO_ID=374955. Retrieved Jan 12th, 2014.

Appendix 6

The 2013 Quota of Exemplary Schools by Districts

Districts	Number
Dongcheng	6
Xicheng	7
Chaoyang	12
Haidian	11
Fengtai	7
Shijingshan	3
Mentougou	3
Fangshan	9
Tongzhou	7
Shunyi	4
Changping	8
Daxing	7
Pinggu	4
Huairou	3
Miyun	4
Yanqing	3
Yanshan	2
Total	100

Source: Beijing Commission of Education 2013, http://zfxxgk.beijing.gov.cn/fgdyna.prinfodetail.prPlanDetailInfo.do?GM_T_CATALOG_INFO/CATA_INFO_ID=374955. Retrieved Jan 12th, 2014.

Appendix 7

Evaluating Indicator System of School Culture

Indicators	Sub-indicators	Rating				
		5	4	3	2	1
Understanding and Ideas of School Culture	Accuracy					
	Integrity					
	Integration					
	Feasibility					
	Consensus					
Spiritual Culture	Core Values					
	Goals of School Development					
	Goals of student development					
	School Motto					
InstitutionalCulture	Legal Awareness					
	Principal Accountability System					
	Check and Balance System					
	Parental Involvement					
	CommunityCooperation					
Operating Culture	Management					
	Curriculum and Instruction					
	Faculty Development					
	Student Development					
	Teacher-Student Interaction					
Environment Culture	Natural Environment					
	School Climate					
	Usage of Space					
HistoricCulture	HistoricCulture					
CulturalCharacteristics	CulturalCharacteristics					

Source: Beijing Commission of Education 2013, http://www.bjedu.gov.cn/Portals/0/fujian/103010.doc. Retrieved Jan 12th,2014.

Appendix 8

Questionnaire on School Culture (for Teachers)

Dear teachers,

This questionnaire has been designed to facilitate a greater understanding of school culture and leadership. It will provide information that may help to improve school management, which in turn may be able to improve the effectiveness of school work. This survey is anonymous as to your school and your personal information and individual responses. The data analysis will not involve any individual analysis and does not have a negative impact on you or your school. Your participation will be of great help. Please read all of the instructions carefully and then fill in each section with your own opinions and feelings, as appropriate.

Thank you very much.

Best wishes,

Beijing Normal University research group

Personal information: Please place a checkmark in the appropriate box

1. Your gender:
   ☐ Male
   ☐ Female

2. Your job duty:
   ☐ Teacher (which subject:__________)
   ☐ Middle leadership

3. Your educational level:
   ☐ Post graduate
   ☐ Bachelor's degree
   ☐ Diploma (3 year higher education but less than a bachelor's degree)
   ☐ Others (please specify)________)

4. Which grade do you teach:
   ☐ First
   ☐ Second
   ☐ Third
   ☐ Fourth
   ☐ Fifth
   ☐ Sixth

5. How many years have you been working as a teacher:
   ☐ 0–5 years
   ☐ 6–15 years
   ☐ more than 15 years

6. Your job title:
   ☐ Junior
   ☐ Middle
   ☐ Senior

There are five options for every statement: "Completely disagree", "Somewhat disagree", "Neutral", "Somewhat agree", "Completely agree". They are coded as 1, 2, 3, 4, 5 respectively. Please place a checkmark on the appropriate number that you feel best reflects your opinion/feeling of the statement, in regard to your school.

Please answer each statement by checking only one number. There is no right or wrong answer.

Cooperation across different parts of the school is actively encouraged.	1	2	3	4	5
Authority is delegated so that teachers can act on their own.	1	2	3	4	5
People from different parts of the school share a common perspective.	1	2	3	4	5
We often have trouble reaching agreement on key issues.	1	2	3	4	5
When disagreements occur, we work hard to achieve "win-win" solutions.	1	2	3	4	5
Most teachers are highly involved in their work.	1	2	3	4	5
It is easy to reach consensus, even on difficult issues.	1	2	3	4	5
There is good alignment of goals across all levels.	1	2	3	4	5
There is continuous financial investment from the school to improve the professional skills of the teachers.	1	2	3	4	5
Information is widely shared so that everyone can get the information he or she needs when it's needed.	1	2	3	4	5
Teamwork is encouraged to get school work done, rather than hierarchy.	1	2	3	4	5
Each teacher can see the relationship between his or her job and the goal of the school.	1	2	3	4	5
The professional skills of teachers are constantly improving.	1	2	3	4	5
Everyone believes that he or she can have a positive impact.	1	2	3	4	5
There is a clear agreement about the right way and the wrong way to do things.	1	2	3	4	5
It is easy to coordinate working across different parts of the school.	1	2	3	4	5
Sufficient cooperation among teachers is taking place in ethical education, teaching, and other areas.	1	2	3	4	5
Planning is ongoing and involves everyone in the process to some degree.	1	2	3	4	5
It is difficult for teachers to cooperate in our school.	1	2	3	4	5
Problems often arise because some teachers do not have the skills necessary to do the job.	1	2	3	4	5
The interests of the parent often get ignored in our decisions.	1	2	3	4	5
I am aware of the principal's educational philosophy and can adjust my own accordingly.	1	2	3	4	5

I understand the school's development goals.	1	2	3	4	5
I understand the school motto's cultural meaning.	1	2	3	4	5
My mode of dress and behavior are consistent with school culture.	1	2	3	4	5
Short-term thinking often compromises our long-term vision.	1	2	3	4	5
I will devote myself to fulfilling the school's development goals.	1	2	3	4	5
Learning is an important objective in our day-to-day work.	1	2	3	4	5
Teachers have a shared vision of what the school will be like in the future.	1	2	3	4	5
Attempts to create change usually meet with resistance.	1	2	3	4	5
Parent comments and recommendations often lead to changes.	1	2	3	4	5
The mode of both teaching and management is very flexible.	1	2	3	4	5
We view failure as an opportunity for learning and improvement.	1	2	3	4	5
Peer schools in the district wish to imitate our development strategy.	1	2	3	4	5
All teachers have a deep understanding of parent wants and needs.	1	2	3	4	5
New and improved ways to do work are continually adopted.	1	2	3	4	5
Leaders set goals that are ambitious, but realistic.	1	2	3	4	5
All school work is conducted with the guidance of school development strategies.	1	2	3	4	5
The leadership has "gone on record" about the objectives we are trying to meet.	1	2	3	4	5
Lots of things "fall between the cracks".	1	2	3	4	5
There is a clear strategy for the future.	1	2	3	4	5
We encourage teachers to have direct contact with parents.	1	2	3	4	5
Leaders continuously track our progress against our stated goals.	1	2	3	4	5
Leaders have a long-term viewpoint.	1	2	3	4	5
The school has an effective strategy for competing with peer schools.	1	2	3	4	5
Our vision creates excitement and motivation for teachers.	1	2	3	4	5
Our strategic direction is unclear to me.	1	2	3	4	5
Innovation and risk taking are encouraged and rewarded.	1	2	3	4	5

Appendix 9

Questionnaire on School Culture (for Principals)

Dear principals,

This questionnaire has been designed to facilitate a greater understanding of school culture and leadership. It will provide information that may help to improve school management, which in turn may be able to improve the effectiveness of school work. This survey is anonymous as to your school and your personal information and individual responses. The data analysis will not involve any individual analysis and does not have a negative impact on you or your school. Your participation will be of great help. Please read all of the instructions carefully and then fill in each section with your own opinions and feelings, as appropriate.

Thank you very much.

Best wishes,

Beijing Normal University research group

Personal information: Please place a checkmark in the appropriate box

1. Your school level:
   ☐ Kindergarten
   ☐ Primary school
   ☐ Junior high school
2. Your gender:
   ☐ Male
   ☐ Female

3. Your educational level:
   ☐ Post graduate
   ☐ Bachelor's degree
   ☐ Diploma (3 year higher education but less than a bachelor's degree)
   ☐ Others (please specify)______

4. How many years have you been a principle:
   ☐ less than 6 years
   ☐ 6–10 years
   ☐ 11–15 years
   ☐ 16–20 years
   ☐ more than 20 years
5. Your job title:
   ☐ Junior
   ☐ Middle
   ☐ Senior

6. How many years have you been working as a teacher:
   ☐ 0–5 years
   ☐ 6–15 years
   ☐ more than 15 years

There are five options for every statement: "Completely disagree", "Somewhat disagree", "Neutral", "Somewhat agree", "Completely agree". They are coded as 1, 2, 3, 4, 5 respectively. Please place a checkmark on the appropriate number that you feel best reflects your opinion/feeling of the statement, in regard to your school.

Please answer each statement by checking only one number. There is no right or wrong answer.

When disagreements occur, we work hard to achieve "win-win" solutions.	1	2	3	4	5
It is easy to reach consensus, even on difficult issues.	1	2	3	4	5
Problems often arise because some teachers do not have the skills necessary to do the job.	1	2	3	4	5
My mode of dress and behavior are consistent with school culture.	1	2	3	4	5
I understand the school's development goals.	1	2	3	4	5
The professional skills of teachers are constantly improving.	1	2	3	4	5
People from different parts of the school share a common perspective.	1	2	3	4	5
I understand the school motto's cultural meaning.	1	2	3	4	5
Most teachers are highly involved in their work.	1	2	3	4	5
Each teacher can see the relationship between his or her job and the goal of the school.	1	2	3	4	5
There is good alignment of goals across all levels.	1	2	3	4	5
It is difficult for teachers to cooperate in our school.	1	2	3	4	5
Everyone believes that he or she can have a positive impact.	1	2	3	4	5
Cooperation across different parts of the school is actively encouraged.	1	2	3	4	5
There is continuous financial investment from the school to improve the professional skills of the teachers.	1	2	3	4	5
Authority is delegated so that teachers can act on their own.	1	2	3	4	5
Information is widely shared so that everyone can get the information he or she needs when it's needed.	1	2	3	4	5
Sufficient cooperation among teachers is taking place in ethical education, teaching, and other areas.	1	2	3	4	5
I am aware of the principal's educational philosophy and can adjust my own accordingly.	1	2	3	4	5
We often have trouble reaching agreement on key issues.	1	2	3	4	5
There is a clear agreement about the right way and the wrong way to do things.	1	2	3	4	5
Planning is ongoing and involves everyone in the process to some degree.	1	2	3	4	5
It is easy to coordinate working across different parts of the school.	1	2	3	4	5
Teamwork is encouraged to get school work done, rather than hierarchy.	1	2	3	4	5
Teachers have a shared vision of what the school will be like in the future.	1	2	3	4	5
Short-term thinking often compromises our long-term vision.	1	2	3	4	5
Parent comments and recommendations often lead to changes.	1	2	3	4	5

All teachers have a deep understanding of parent wants and needs.	1	2	3	4	5
Our vision creates excitement and motivation for teachers.	1	2	3	4	5
Our strategic direction is unclear to me.	1	2	3	4	5
I will devote myself to fulfilling the school's development goals.	1	2	3	4	5
New and improved ways to do work are continually adopted.	1	2	3	4	5
Attempts to create change usually meet with resistance.	1	2	3	4	5
The mode of both teaching and management is very flexible.	1	2	3	4	5
Innovation and risk taking are encouraged and rewarded.	1	2	3	4	5
Leaders set goals that are ambitious, but realistic.	1	2	3	4	5
There is a clear strategy for the future.	1	2	3	4	5
The interests of the parent often get ignored in our decisions.	1	2	3	4	5
Leaders have a long-term viewpoint.	1	2	3	4	5
All school work is conducted with the guidance of school development strategies.	1	2	3	4	5
We encourage teachers to have direct contact with parents.	1	2	3	4	5
The school has an effective strategy for competing with peer schools.	1	2	3	4	5
Lots of things "fall between the cracks".	1	2	3	4	5
We view failure as an opportunity for learning and improvement.	1	2	3	4	5
Learning is an important objective in our day-to-day work.	1	2	3	4	5
Peer schools in the district wish to imitate our development strategy.	1	2	3	4	5
Leaders continuously track our progress against our stated goals.	1	2	3	4	5
The leadership has "gone on record" about the objectives we are trying to meet.	1	2	3	4	5

Appendix 10

Instrument: Items by Index and Trait

The Denison Organizational Culture Survey (with the permission from Denison Consulting) was employed as the major data generation instrument in this study. Minor revisions were made to the original survey in order to make the translation more closely fit the Chinese-language context and to better suit the school context in China.

Trait	Index	Item
Involvement	Empowerment	1. Most teachers are highly involved in their work. 2. Everyone believes that he or she can have a positive impact. 3. Planning is ongoing and involves everyone in the process to some degree. 4. Information is widely shared so that everyone can get the information he or she needs when it's needed.
	Team Orientation	5. Cooperation across different parts of the school is actively encouraged. 6. Sufficient cooperation among teachers is taking place in ethical education, teaching, and other areas. 7. Teamwork is encouraged to get school work done, rather than hierarchy. 8. Each teacher can see the relationship between his or her job and the goal of the school.
	Capability Development	9. Authority is delegated so that teachers can act on their own. 10. The professional skills of teachers are constantly improving.

		11. There is continuous financial investment from the school to improve the professional skills of the teachers. 12. Problems often arise because some teachers do not have the skills necessary to do the job. (Reversed Scale)
Consistency	Core Values	13. I am aware of the principal's educational philosophy and can adjust my own accordingly. 14. I understand the school's development goals. 15. I understand the school motto's cultural meaning. 16. My mode of dress and behavior are consistent with school culture.
	Agreement	17. When disagreements occur, we work hard to achieve "win-win" solutions. 18. It is easy to reach consensus, even on difficult issues. 19. There is a clear agreement about the right way and the wrong way to do things. 20. We often have trouble reaching agreement on key issues. (Reversed Scale)
	Coordination and Integration	21. People from different parts of the school share a common perspective. 22. It is easy to coordinate working across different parts of the school. 23. It is difficult for teachers to cooperate in our school. (Reversed Scale) 24. There is good alignment of goals across all levels.
Adaptability	Creating Change	25. The mode of both teaching and management is very flexible. 26. The school has an effective strategy for competing with peer schools. 27. New and improved ways to do work are continually adopted. 28. Attempts to create change usually meet with resistance. (Reversed Scale)

	Customer Focus	29. Parent comments and recommendations often lead to changes. 30. All teachers have a deep understanding of parent wants and needs. 31. The interests of the parent often get ignored in our decisions. (Reversed Scale) 32. We encourage teachers to have direct contact with parents.
	Organizational Learning	33. We view failure as an opportunity for learning and improvement. 34. Innovation and risk taking are encouraged and rewarded. 35. Lots of things "fall between the cracks". (Reversed Scale) 36. Learning is an important objective in our day-to-day work.
Mission	Strategic Direction & Intent	37. Peer schools in the district wish to imitate our development strategy. 38. All school work is conducted with the guidance of school development strategies. 39. There is a clear strategy for the future. 40. Our strategic direction is unclear to me. (Reversed Scale)
	Goals & Objectives	41. Leaders set goals that are ambitious, but realistic. 42. The leadership has "gone on record" about the objectives we are trying to meet. 43. Leaders continuously track our progress against our stated goals. 44. I will devote myself to fulfilling the school's development goals.
	Vision	45. Teachers have a shared vision of what the school will be like in the future. 46. Leaders have a long-term viewpoint. 47. Short-term thinking often compromises our long-term vision. (Reversed Scale) 48. Our vision creates excitement and motivation for teachers.

Lightning Source UK Ltd.
Milton Keynes UK
UKOW06n0610081215

264273UK00001B/63/P

9 788793 102668